THE COMPLETE YACHT SECURITY HANDBOOK

THE COMPLETE YACHT SECURITY HANDBOOK

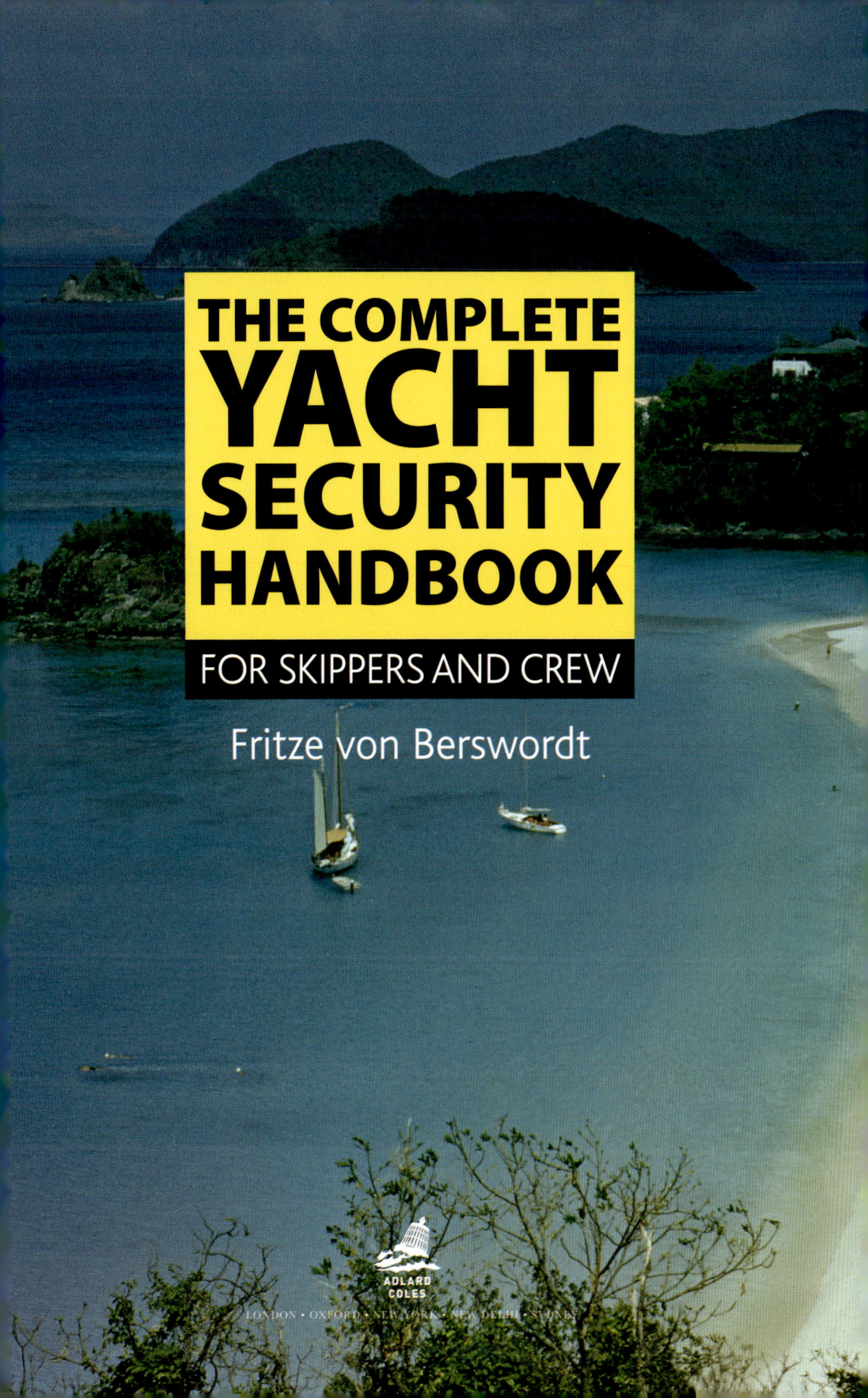

FOR SKIPPERS AND CREW

Fritze von Berswordt

ADLARD COLES
Bloomsbury Publishing Plc
50 Bedford Square, London, WC1B 3DP, UK
www.adlardcoles.com

BLOOMSBURY, ADLARD COLES and the Adlard Coles logo are trademarks of
Bloomsbury Publishing Plc

First published in Great Britain 2018

Copyright © Fritze von Berswordt, 2018

Fritze von Berswordt has asserted his right under the Copyright, Designs and Patents Act, 1988, to be identified as Author of this work

All rights reserved. No part of this publication may be reproduced or transmitted in any form or by any means, electronic or mechanical, including photocopying, recording, or any information storage or retrieval system, without prior permission in writing from the publishers

Bloomsbury Publishing Plc does not have any control over, or responsibility for, any third-party websites referred to or in this book. All internet addresses given in this book were correct at the time of going to press. The author and publisher regret any inconvenience caused if addresses have changed or sites have ceased to exist, but can accept no responsibility for any such changes

A catalogue record for this book is available from the British Library

Library of Congress Cataloguing-in-Publication data has been applied for

ISBN: PB: 978-1-4729-5167-0; ePub: 978-1-4729-5168-7; ePDF 978-1-4729-5169-4

2 4 6 8 10 9 7 5 3 1

Designed by CE Marketing
Printed and bound in China by Toppan Leefung Printing

MIX
Paper from responsible sources
FSC® C104723

To find out more about our authors and books visit www.bloomsbury.com and sign up for our newsletters

> No responsibility for loss or damage caused to any individual, company or property by acting on or refraining from action as a result of material in this work is accepted by Bloomsbury or the author. Confrontations with criminals are dangerous and dynamic events and it is not possible to predict how attackers will react. Readers are advised not to take any steps that could bring harm to themselves or others.

CONTENTS

Introduction   **6**

Part I: Background: Acts of Piracy and Attackers   **8**
Piracy: understanding the threat   10
Criminal attacks on yachts   12
Crime dynamics: an introduction to avoiding and repelling criminal attacks   17
Offenders, modes of action and reaction to resistance   28
Physical conflict and self-defence   35

Part II: Planning Your Cruise in Uncertain Regions   **36**
Security balance: a simple method for assessment and planning   38
A clear picture on risk: cruising area and yacht situation   39
Optimising your cruise: effective risk reduction   46

Part III: Preparation: Vessel, Equipment and Crew   **48**
Ship layouts and security   50
Security equipment on board   55
Behaviour and capabilities: preparing the crew   101

Part IV: Security in Action: Avoidance and Deflection of Criminal Attacks   **106**
Safety in the marina   108
Mooring buoys and solitary bays   112
Sailing in pirate-plagued waters   130
Sailing in convoy   163

Appendices   **182**
Items of a crime report   182
Emergency communication table   183
Glossary   **184**
About the author   **187**
Reference   **188**
Index   **190**

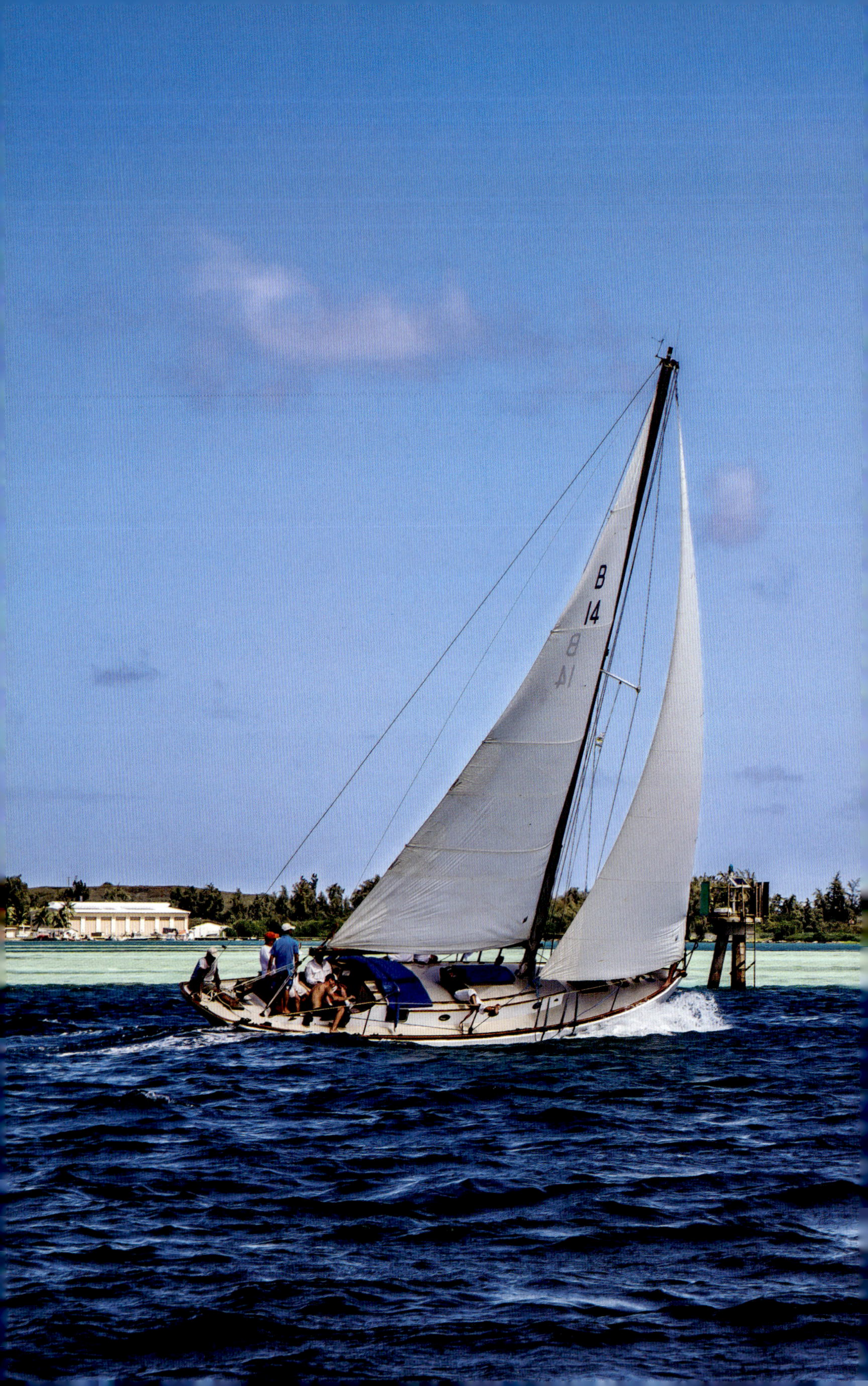
B
14

INTRODUCTION

While circumnavigating our beautiful planet on our trusted catamaran *Alytes*, our family of three on many occasions chose routes that were off the beaten track. This meant that whenever we prepared to set sail for a region rarely visited or even avoided by fellow sailors we had to make some serious decisions about security.

Unfortunately, the background information required for such decisions was largely missing. What's more, we were surprised by how diverse the opinions and the assumed knowledge about criminal attacks aimed at yachts were – diversity did not really help when we were trying to make informed choices.

In response, I decided to research the matter and put together a set of reliable data combined with a collection of strategies and tactics for charter sailors, long-term cruisers and yacht owners who keep their boats outside the waters of northern Europe or the USA.

To do this, I researched more than 250 attacks against cruising yachts between 2011 and 2017. This information, together with conclusions drawn by recent criminological research and reports from real-life experiences, meant that for the first time questions concerning equipment, weapons and resistance tactics could be answered, based on solid information.

Just a few simple methods and preparations are needed to avoid criminal attacks altogether, or else to successfully repel and react to most types of criminals should a situation arise. These easy-to-implement strategies will help owners, skippers and crew members to develop confidence and focus on what counts when going for a sail: making great discoveries, taking part in challenging sports and relaxing while watching the sunset at anchor in beautiful bays.

The following chapters aim to help sailors to get a realistic feel for the potential criminal threats. They cover topics such as effective assessment of cruising areas and the ship's situation, technical equipment, and the crew's preparation for an incident. Real-life examples, technologies and procedures for securing a yacht form the basis for the implementation of sound security on board a vessel.

The result is a fact-based handbook that should empower all cruising yacht skippers to sail to places they did not dare to visit before and to prepare them for destinations they formerly visited unprepared. Whether you plan to sail in Latin America, the Caribbean, the Med or any other place in the world, I hope this book will inspire confidence rather than fear.

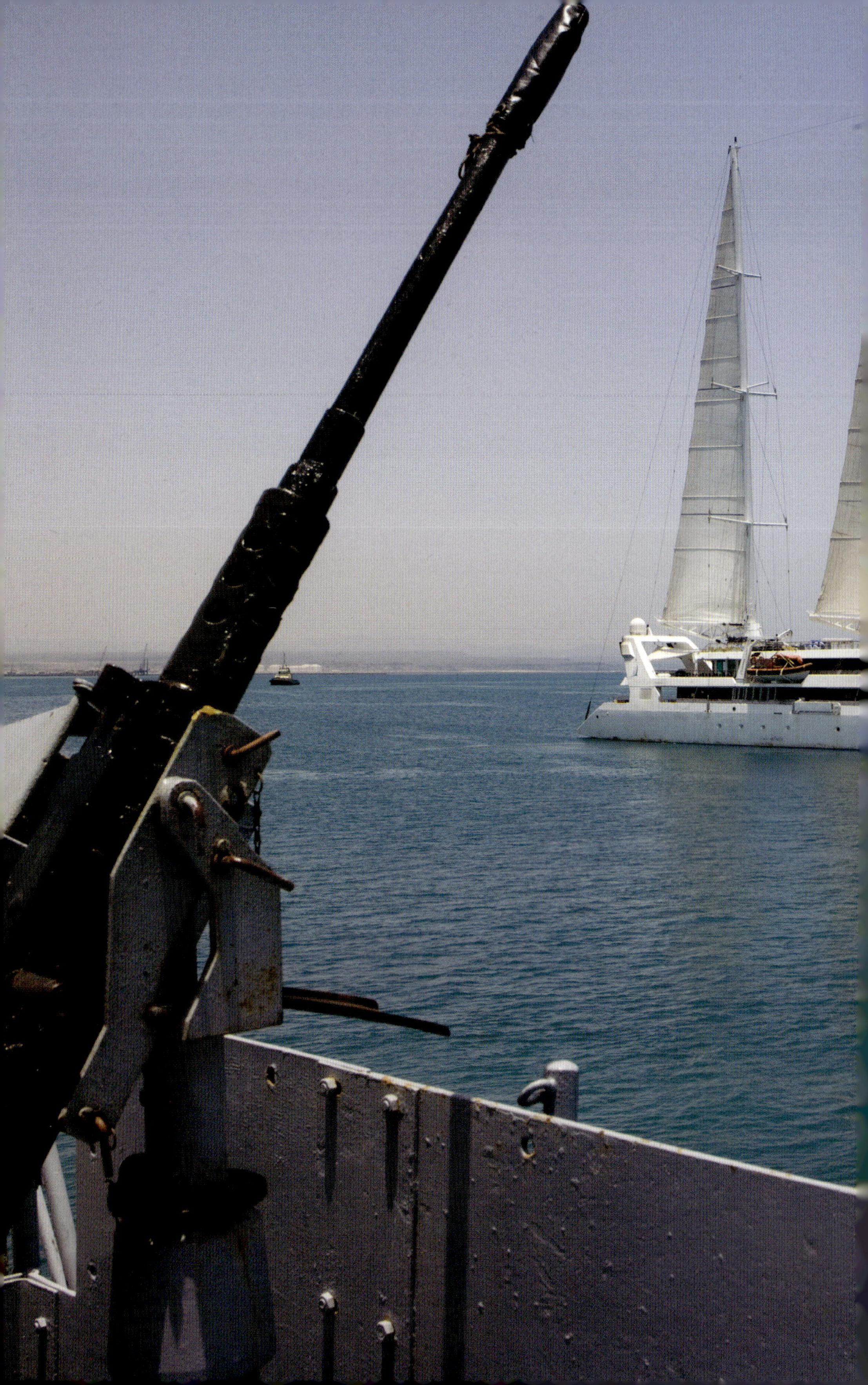

Part I
BACKGROUND: Acts of Piracy and Attackers

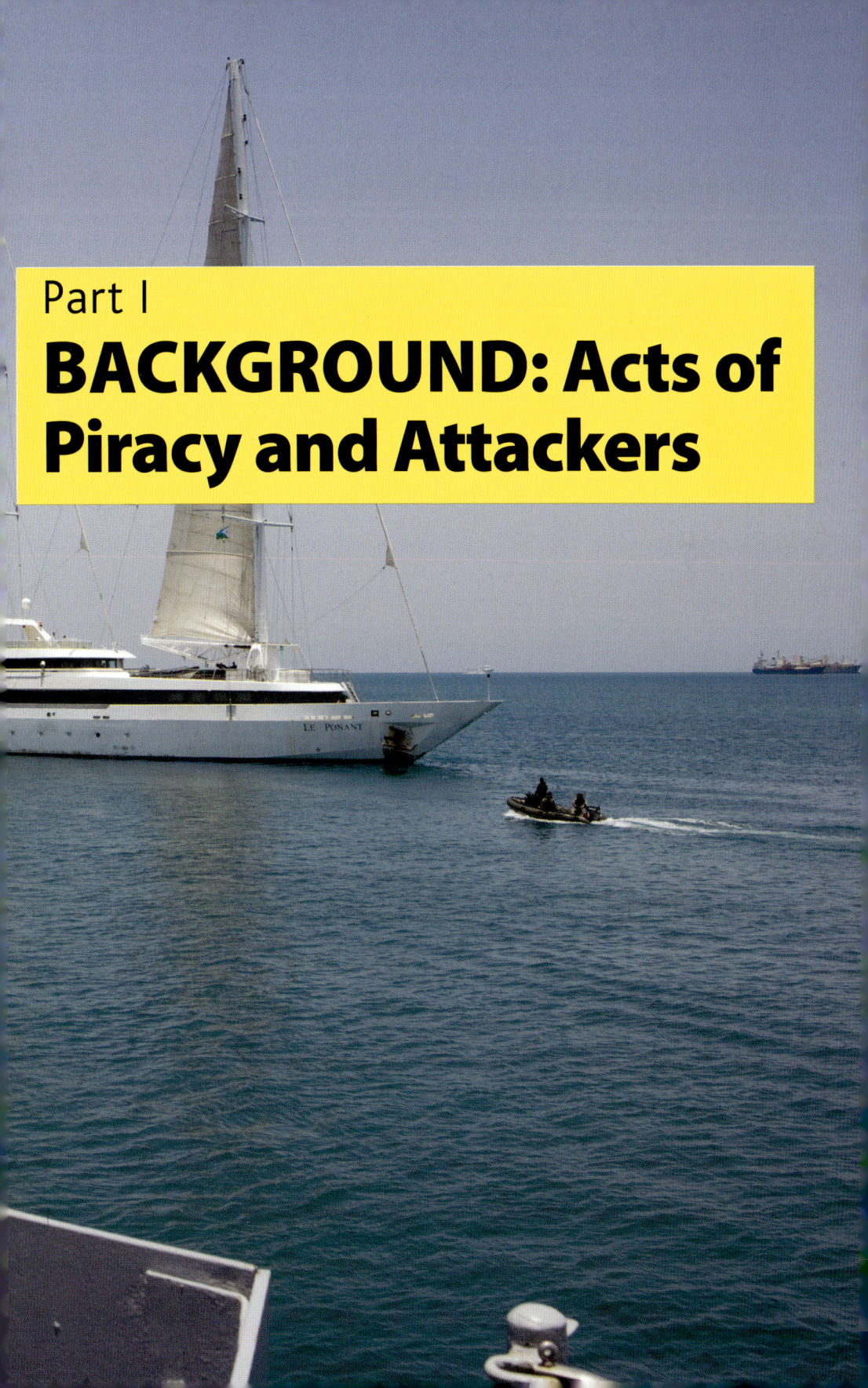

PIRACY: UNDERSTANDING THE THREAT

Understanding the true threat of piracy will help mariners to enjoy a relatively carefree journey to the remotest regions of our planet. In order to do this, the ability to distinguish unsubstantiated stories and reports concerning crimes against commercial shipping from events relevant to cruisers is an important basic step.

My study of more than 250 attacks on yachts around the globe has provided fact-based insights about the type of attacks and the primary modes of action used by criminals attacking cruising yachts. Analysing the dynamics of many assaults and robberies revealed a surprise: resistance to attackers was a good tactic in many situations. Resisting crews were more successful at protecting their property and they did not sustain more (or more severe) injuries than passive crews.

However, obtaining and keeping the initiative in such confrontations proved to be vital for a successful defence – a conclusion that means we should focus on preparation before and actions taken during a confrontation.

As long as mankind has sailed the oceans, pirates have tried to gain their share of the wealth promised by those excellent vessels and their abundant cargo. But what exactly is an act of piracy and how is this relevant for the cruising community?

The basic definition of piracy is any unlawful attack on a ship in international waters with weapons from a distance, or the unlawful and uninvited boarding of a ship in these waters.

FIG 1.1: Solitude – a dream that can leave you vulnerable. (Photo: Getty)

By contrast, all attacks within the national waters of any countries are usually defined as 'normal crime'.

When you are planning to go sailing, you want to be sure that you will return safely. Depending on your personal ideas about exploring, you may also want be certain that the benefits of your travels exceed the risks involved. As a competent sailor, it is likely that you will be well prepared, based on sound and comprehensive information, for the challenges that may be posed by weather, navigation and your vessel. Evaluating the risks of piracy, however, does not seem to be so easy, given that most of the media that informs our opinions about geographic regions and their inhabitants do not differentiate on a level that is helpful to yachtsmen. The media's focus is very much on commercial shipping and usually on audacious attacks made by small bands of rifle-bearing, grappling-hook-swinging boatsmen against huge tankers or container ships.

Consequently, we often shun regions where commercial shipping is targeted and sail other, possibly more dangerous ones, unprepared. Let's look at two prominent examples.

The Strait of Malacca is still regarded as extremely dangerous in public opinion as well as the yachting community. It has a history of violent and undifferentiating attacks on boaters of all kinds, and these attacks are still going on today. However, a closer look reveals that the attacks are now exclusively boardings of large commercial ships anchored off Singapore or Malay harbours. Yachts have not been attacked for many years: we are simply not the target of the criminals active in this area. Despite this situation, though, many skippers do not dare to navigate this convenient passage from Australia to Thailand.

On the other hand, thousands of boats visit the beautiful islands of the southern Caribbean. There is not much news of piracy in these waters, unless, that is, we dig deeper into the information systems specifically made for bluewater sailors. Consequently, many crews leave their dinghies unsecured and their hatches unlocked in the marvellous tropical nights. For some of these crews, the visit to paradise will later be overshadowed by the sobering discovery of a tender stolen from their transom or – in the worst case – waking up to the uncomfortable itch of a machete being held to their throats.

What we are essentially saying is that it is important to consider piracy and maritime crime from the specific perspective of a yachting crew, rather than relying on the somewhat broad view that public media offers.

Instead, we should be asking more questions. What are the true risks for my type of boat and crew? What types of attacks are being carried out against pleasure vessels and by whom? How should I act and how will they react to my actions? With this in in mind, this book aims to help you to understand piracy and develop a realistic and relevant picture of the regions that you are considering visiting, which in turn should enable you to sail fearlessly to the remotest places feeling well prepared, confident and ready to relax.

CRIMINAL ATTACKS ON YACHTS

Yachts cruising oceans and lingering in cosy marinas are confronted with eight different types of attacks: theft, burglary, robbery, hijacking, kidnapping, murder, terrorism and acts of war. The offenders' motivation for the first five is mostly the same: steal the largest booty with the smallest risk. Murder of visiting yachties is mostly an accidental by-product of a robbery gone wrong. Terrorist attacks differ substantially: although terrorists may be interested in valuables or ransom to fund their activities, another strong motive is media coverage to support political, religious or other types of ideological goals.

To more easily understand the offenders' tactics, attacks are grouped into two major groups, as shown in Fig. 1.2. Based upon their most prominent features, we will call these groups 'silent attacks' and 'controlling attacks'.

Silent attacks	Controlling attacks
Attacker's strategy Speed and stealth	**Attacker's strategy** Overpower and control targets
Typical silent attacks Theft Burglary	**Typical controlling attacks** Robbery Hijack Abduction Murder Terroristic attack

FIG 1.2: All attacks against yachts can be allocated to one of two groups.

During a **silent attack**, the offenders will try to avoid any confrontation with crew or potential witnesses. They will, therefore, attempt to snatch unsecured valuables or covertly breach entry barriers, grab their booty and retreat unnoticed or at least unidentified. Theft and burglary are grouped in this category.

The key feature defining a **controlling attack**, by contrast, is the attempt to overpower the targets. A confrontation – dreaded by a silent attacker – is a vital part of any such attack. Robbery, hijack, kidnapping, murder and most terroristic attacks are categorised as controlling attacks.

It is important to bear in mind that an escalation from silent attacks to controlling ones is possible during the dynamic course of a criminal assault. A burglary planned to be silent may develop to a robbery if the crime is discovered and the criminal is not

BACKGROUND 13

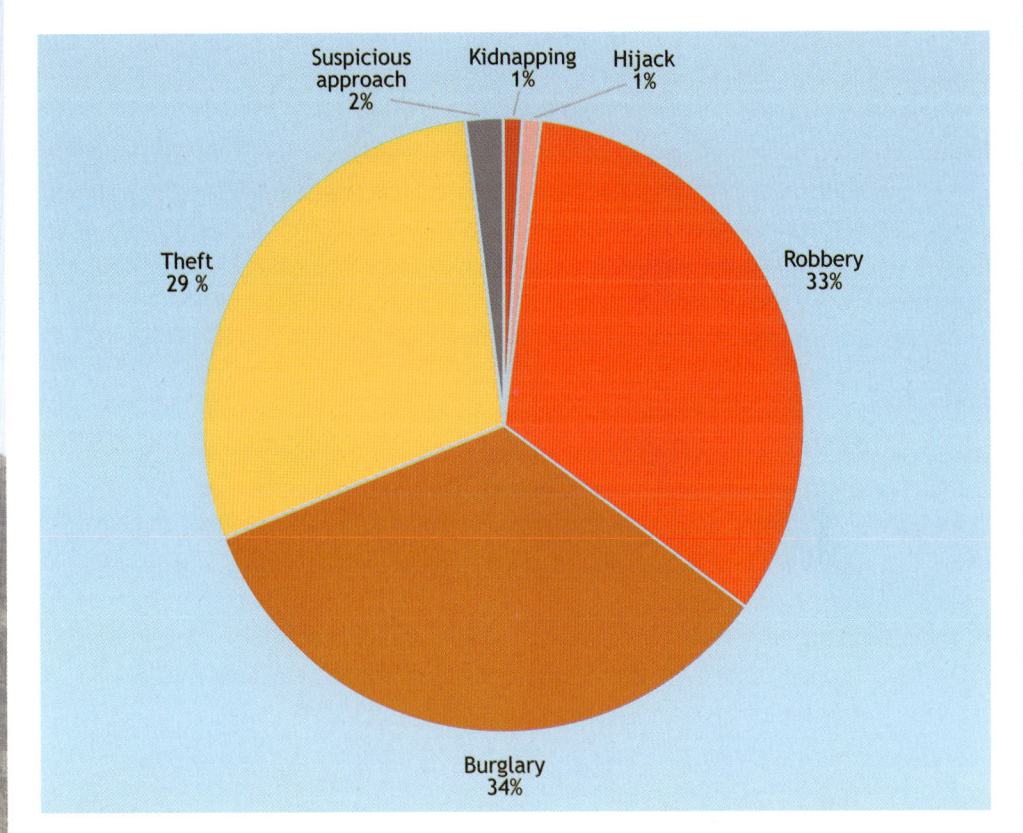

FIG 1.3: *Share of crime types reported against cruising yachts in cruising destinations, including all vessels – manned and unmanned. (Source: 'Yacht Security 2017')*

immediately 'chased away' or controlled by the defending crew. Escalations within a group are also possible, whereby robberies may occasionally end in mayhem or even murder.

Fig. 1.3 shows the share of crimes against cruising yachts between 2011 and 2017 in global cruising regions. The data from the source for this information does not cover petty crime and burglaries against resident yachts in the ports of the northern hemisphere, such as the USA, Japan and Europe. Consequently, the incidence of theft and burglary on a worldwide scale is expected to be a lot bigger.

The attacks can be encountered in the marina, at anchor and – with the exception of theft and burglary – at sea. We will not differentiate the attacks according to the yacht's situation at this time – situation-specific traits of each attack are explored later on.

Theft

Theft is a silent attack and is considered one of the least dangerous criminal activities.

On pleasure crafts in cruising regions, theft targets material assets that the thief can steal from easily accessible and unprotected areas, typically the transom, deck or a marina dock.

Snatching a dinghy and outboard tied to the yacht or a dock is quite common in many parts of the world.

If you take a look at the situation in the West, theft is a massive problem for mariners. For example, US insurance companies reported approximately 5,000 watercraft stolen in 2015. One in five of the identified types were personal watercrafts (jet-skis). However, 52 sailboats, 181 motor yachts (cruisers) and a little more than a thousand runabouts and utility boats were stolen, too. During the same year, roughly 1,350 outboard engines were reported stolen in Germany, and UK records registered more than 3,000 outboards as having 'gone missing' as of November 2016. Regardless of the theft incidents, however, all these countries are considered 'safe' among the cruising community.

Despite being considered harmless, theft can of course cause substantial financial losses as well as great annoyance. Depending on the cruising area, thieves target and cut your halyards and remove solar panels or radar systems. Especially in remote areas, these critical items will be very difficult to replace in a timely manner. Even a stolen dinghy fuel tank can cause a lot of trouble in sparsely inhabited, marina-free areas of the world, if there is no replacement on board.

Thieves aim to secure large gains without having to overcome complex barriers and without being detected or even identified. Therefore, a direct attack against any person is not their goal. However, theft can escalate in two ways:

- If a thief discovers attractive items inside the vessel while roaming its deck for booty, he may decide to break into the cabin, thus becoming a burglar. Open hatches or unlocked saloon doors could also tempt a thief to enter your boat.
- If a thief is surprised by the crew and his escape route is blocked then he may attack. Usually the main goal of such an attack would be to ensure an unidentified getaway. In rare cases, a thief may become a robber if he has the chance (and personality) to overpower the crew during a theft attempt.

In most cases, though, thieves will try to escape the moment they are discovered or even confronted.

Burglary

In contrast to a simple theft, the burglar will try to gain access to areas or rooms that are barred by more-or-less effective barriers (such as doors or hatches). Burglary is a silent attack: a burglar's main tactic aims to remain unnoticed and unidentified while entering, looting and exiting the target.

Burglary usually causes more damage than theft. On the one hand, the burglar has access to areas that are often used to store more-valuable assets, while on the other hand the break and entry causes damage to vital boat equipment, such as hatches and companionway doors. Such damage is difficult to repair properly without the correct tools and training. Given that leaving a safe port or anchorage with damaged hatches is not advisable, prolonged periods of repairs or waiting for spare parts may follow a burglary. Such periods are aggravated in remote areas with weak service infrastructure.

Burglars usually have a very different personality to thieves. They judge themselves as being capable of overcoming entry barriers, bring tools and are generally more prepared and experienced in criminal activities. They are also usually more willing to take risks. By entering indoor areas, they are in danger of surprising people inside or being surprised by a crew member returning from a shore excursion. Whenever a burglar decides to confront someone rather than escaping, the burglary may escalate to become a robbery.

However, as long as an escape route is available, burglars are more likely to run than to fight since they want to avoid being detected or confronted.

Robbery

As with thieves and burglars, the robber's primary goal is to get hold of as many valuables as possible. In rare cases, this may include the vessel itself.

The fundamental difference between robbery, theft and burglary is the robber's tactic, which aims to control crew and passengers. Overpowering the targets is a key aspect of his crime, and the robber will try to achieve control by threatening or executing physical violence against everybody aboard. This means that they come prepared for a violent confrontation, and are ready and willing to threaten and injure other people with or without the use of weapons. Weapons – especially firearms – will help a small group of robbers or single individuals to successfully control a much larger group of physically stronger opponents.

Robbers take immense risks. Unless they are able to recon the target boat and its crew, they enter a confrontation with very little knowledge about the location and relevant characteristics of the opposing force (such as their number, determination and armament). Accordingly, they are usually experienced criminals, most likely with a history of confrontation and violence. This makes them very dangerous attackers.

Both material and psychological damages caused by a successful robbery are high. The criminals are free to choose their booty once the targets have been effectively controlled. In many cases, robbers will threaten their victims to get them to reveal hiding places and open strongboxes to obtain especially valuable assets, and even if victims surrender early, this aggressive approach can lead to casualties. What's more, the confrontation and loss of control experienced by the victims can lead to trauma and subsequent psychological conditions after the event. In some cases, female crew members are subject to sexual harassment or – very seldom – rape during a controlling situation.

A robbery provides the potential for injuries (both physical and psychological) and death. Crews who have been able to successfully fend off robbers were those who acted swiftly, were well armed, and who used their initiative.

Abduction

An abduction aims for the long-term control of the attacked targets. However, in this instance control needs to be maintained for longer, while victims are transported to shore and during the following hostage phase. Abductions are thus categorised as controlling attacks.

The distance from the crime scene to the long-term prison can range from a few thousand metres up to several thousand kilometres. In many cases, kidnappers have some accomplices or even a larger group of supporters at hand in the region where they plan to hold the hostages.

Abductions are usually crimes of determined, prepared attackers who are willing to resort to violence. In order to carry out a kidnapping, criminals have to build up and maintain an infrastructure that can host a crew for a longer time. Consequently, they are considered to be organised criminals or terrorists. As such, assailants are generally trained in the use of weapons, assault tactics and carrying out and handling violent confrontations. They will most likely act according to a well-laid plan that includes routes of retreat with or without the hostages.

Both the psychological and physical damage of a successful abduction can be tremendous. Whenever kidnappers are successful in controlling their targets and transporting them to the intended prison, an unassisted escape is extremely unlikely. A prolonged period of hardship, insecurity and abuse will be the most likely scenario following this stage.

An abduction is an extremely dangerous situation. Injuries, mayhem, rape and murder are possible both at the time when the attackers try to gain control as well as after the successful transportation and imprisonment. It is thus very risky to resist well-armed and well-organised kidnappers with a terrorist background.

Murder

In most attacks against yachts, murder is not the original goal of the action. Instead, it may be the result of a failed attempt to control the crew for a robbery or abduction. A special case is murders that rid the attacker of witnesses or relieve him of the burden in a poorly planned or poorly executed abduction. Unfortunately, killing hostages when ransom is not paid in time is a recurring pattern for crimes conducted by terrorist groups. The groups' primary goal – media coverage for their cause – is still achieved even if money is not handed over.

In some rare cases, assassination of the victim was the primary goal of the offence. In these instances, the boaters were either residing in a fixed position for a long time or the police suspect fellow crew members of the lethal attack.

Terrorism and acts of war

Terrorists usually aim at murder, mayhem or abduction as a means of supporting their political or religious goals. This makes a terrorist attack fundamentally different because the typical rationale of a property crime does not apply.

Terrorists approach in medium to large groups and usually carry heavy armament. Deceptive tactics, such as dressing up as police officers or simply claiming to be an official detachment, have been used in past attacks. Terrorists are very determined and psychologically hardened by ideological or religious indoctrination. Some individuals could be driven by a philosophy of martyrdom. These subjects are unlikely to behave rationally from a Western perspective. Considerations of their own survival may not be part of their action plan.

The typical yacht crew will thus face major challenges while repelling a terrorist attack, so escape can be the best, albeit risky, option. In at least one case – a nocturnal attack on a marina – some of the victims managed to jump into the water and escape in the dark by dodging between boats.

Purposeful or accidental attacks carried out by parties in a regular or civil war are a major risk in war zones. Due to the methods and weapons normally used, such military attacks are likely to prove fatal for a civilian target.

This book does not cover the dynamics of a terrorist attack or an attack conducted by regular armed forces aimed at killing or kidnapping; instead, we recommend at all costs avoiding a visit to areas where there is a risk of meeting such offenders. Except for the rare cases in which a war situation arises just as a crew arrives, such regions are well known and can be given a wide berth. Please see the 'Planning Your Cruise in Uncertain Regions' section (see page 36–47) for further details concerning the identification of such regions.

CRIME DYNAMICS: AN INTRODUCTION TO AVOIDING AND REPELLING CRIMINAL ATTACKS

Criminal acts create dynamic situations, the outcomes of which depend on many factors. The most relevant of these will be covered here. When comparing different crime types, there is a pattern that seems to be present in many criminal attacks.

The motivations, preparations and tactics for a crime's execution are quite similar within the 'silent' and 'controlling' attack groups shown in Figs. 1.4 and 1.5. However, there are relevant differences between these groups.

As outlined in 'A clear picture on risk: cruising areas and yacht situation' (see page 39–45), we have to primarily expect silent attacks in safer cruising areas (Risk Levels I and II), such as

Northern Europe, larger parts of the Mediterranean and the safer places in the Caribbean. Accordingly, sailors can relax and enjoy themselves in these places if the crew and ship are prepared to avoid or repel theft and burglary. As soon as a crew decides to sail the waters of higher-risk areas (Risk Level III), however, they should be ready to cope with occasional controlling attacks.

Trying to see things from the offenders' perspective is a good start when thinking about some very basic aspects concerning avoidance and repelling tactics.

Silent attacks

Most thieves and burglars are in it for the easy, low-risk booty. They want to be quick and stealthy during the crime. Any activity or device that aims at disturbing that scheme (speed and stealth) will help to protect against silent attacks. Consequently, the most important measures to guard against such attackers are to:

- Reduce attractiveness and minimise opportunity: hide all valuable items.
- Reduce movability: secure all items that you cannot hide effectively, using conspicuous means that discourage criminal observers.
- Hinder access: fit locks and other barriers that are able to withstand the tools and methods usually employed by criminals in the region visited. Never leave your boat unlocked while away or asleep.
- Detect and deter: install effective alarm systems that will detect thieves and burglars before they cause damage by tampering with or overpowering entry barriers.

Silent attacks

Triggers and preconditions
- Attacker assumes or sees attractive booty
- Attacker is confident of being able to access booty
- Attacker is confident they will remain stealthy/unidentified (not seen, not heard, speedy)

Sequence (stealth, speed)
- Select target
- Board and breach entry barriers (if existent)
- Retreat and escape

Defensive measures
- Minimise attractiveness, clear deck
- Install conspicuous deterrents (deter and alarm)
- Install conspicuous entry barriers and secure visible valuables (deter and slow down)
- Determined repelling (with initiative)

FIG 1.4: Silent attacks – preconditions, typical sequence and defensive measures.

Countermeasures that can be seen from a distance will help to prevent the crime altogether. A clean deck, strong and visible locks and obvious alarm systems or signs warning of such systems will reduce any offender's motivation. The countermeasures also work during the attack itself. The longer a thief takes to get hold of valuables, the higher the chances that he will abort and retreat. An alarm system with a howling siren and bright lights exposing the offender will disorient and repel many silent attackers.

By contrast, all those yachts spending the night with a load of unsecured jerry cans on deck, a loosely bound surfboard at the stanchions, the dinghy bobbing peacefully in the water and all hatches wide open will attract criminals as efficiently as an anchor light lures in tropical insects.

Controlling attacks

Unlike silent attackers, robbers and kidnappers try to gain control over crew and passengers by threats and violence. However, just like silent attackers, they try to avoid being seen by external witnesses, let alone security forces. They thus prefer to attack crews who appear to be easily surprised and quickly controlled as well as being isolated, weak and not very alert. Consequently, countermeasures should focus on early detection as well as conveying the impression of alertness and strength:

- Appear strong (this includes making friends ashore).
- Avoid surprises.
- Avoid being boarded.
- Call for assistance.
- When you have the initiative, resist and repel.

Controlling attacks

Triggers and preconditions
- Attacker assumes or sees attractive booty
- Attacker is confident of being able to overpower crew
- Attacker is confident of being able to escape

Sequence (surprise, overpower)
- Select target/ambush
- Surprise/offensive approach
- Board
- Retreat and escape

Defensive measures
- Stay stealthy
- Destroy element of surprise, signal strength and determination
- Signal for help and call in support, make noise
- Determined repelling (with initiative)

FIG 1.5: Controlling attacks – preconditions, typical sequence and defensive measures.

Depending on their level of competence, robbers and kidnappers plan their attacks, taking into account risks and the chances of success to a varying degree. If a crew succeeds in presenting a ship that looks difficult to board and is staffed with a crew that is vigilant as well as willing and able to defend, an attack can be avoided altogether. Offenders will be more likely to hesitate before launching an assault that may involve a higher risk of early detection, a lack of surprise, tough resistance and a lot of commotion. All of these elements make a successful robbery less likely.

The general tactics of attackers and consequently defenders can be simplified a little further, as can be seen in Fig. 1.6.

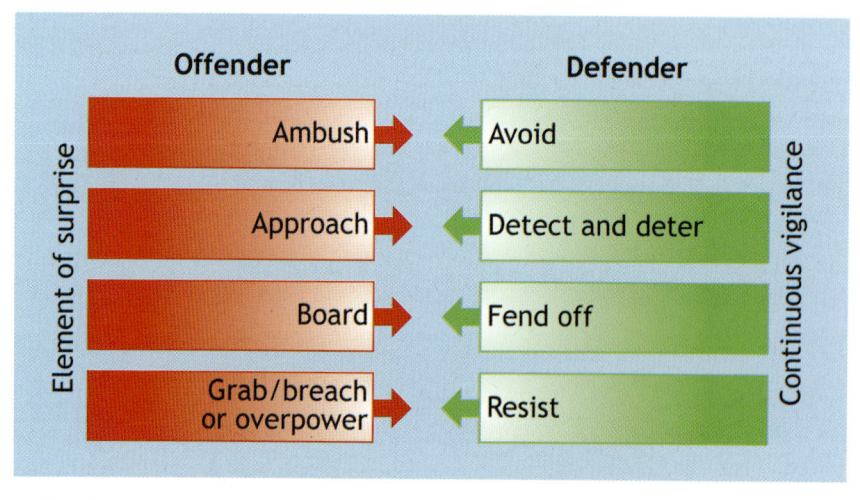

Offender	Defender
Ambush →	← Avoid
Approach →	← Detect and deter
Board →	← Fend off
Grab/breach or overpower →	← Resist

(Element of surprise / Continuous vigilance)

FIG 1.6: Attackers' actions and defenders' counter-actions (simplified).

You might have noticed that the simplification raises some minor issues: thieves and burglars do not ambush (they may merely ferret out your boat) and resistance against gun-wielding robbers on board may be a bad idea even if the crew is armed in a similar way. Nevertheless, we will resort to this scheme in the following chapters so that we can assign actions or equipment to the respective phase of an attack.

Do not be misled in the assumption that your ship has to look like a fortress each night at anchor with someone patrolling the lifelines. However, hard-to-climb sides, an effective light on deck during the night and very firm behaviour towards anyone approaching your vessel after dark may help to reduce any risks.

In seedier environments, you may feel much more relaxed if you have a simple plan of action that you can put into place should there be a nightly boarding. This would include actions that will alert the coastal community, repel boarders and generally withstand any attempt to be controlled by intruders. The more effectively you are able to demotivate boarders from performing an attack by using conspicuous equipment, the more likely it is that you will avoid the confrontation in the first place.

BACKGROUND **21**

Now, what about the moment you hear a boat touching your sides in the night or, worse, footsteps on your deck? Let us have a look at data from more than 250 attacks on yachts between 2011 and 2017, focusing on ones that involved a crew being on board, so we can see which tactics worked best.

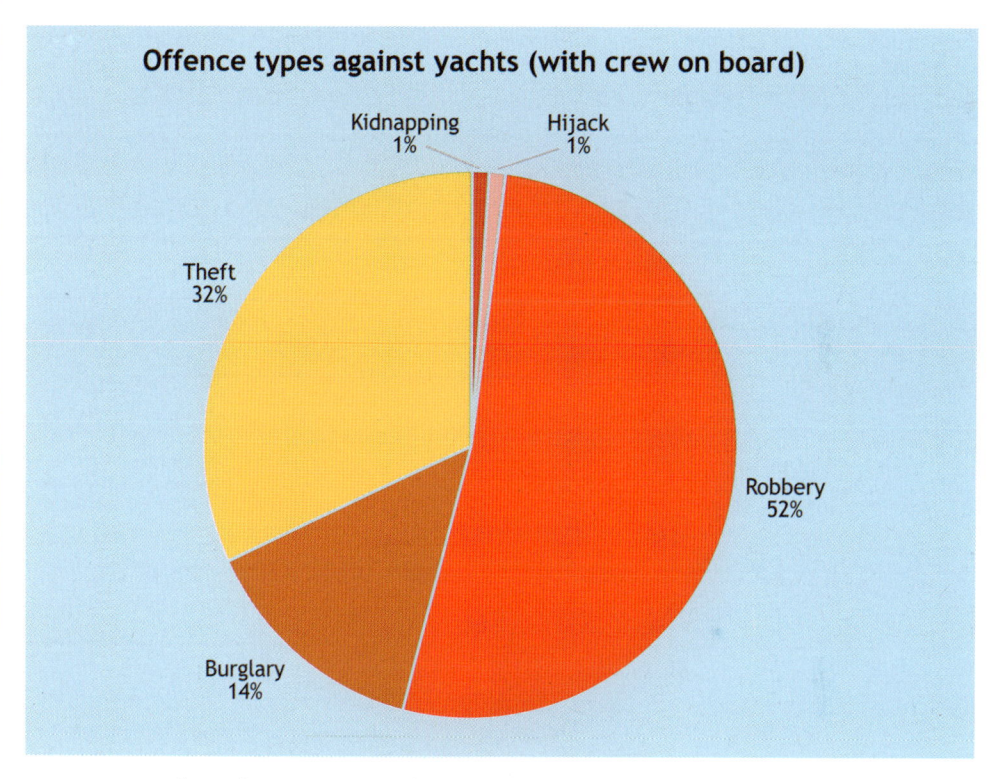

Offence types against yachts (with crew on board)

- Kidnapping 1%
- Hijack 1%
- Theft 32%
- Robbery 52%
- Burglary 14%

FIG 1.7: *Type of attacks against crewed yachts. (Source: 'Yacht Security 2017')*

When looking at the data in Fig. 1.7 you can see that just slightly more than half of reported attacks (robbery, kidnapping and hijack) included attempts to actually control the crew. The remaining half were intended as silent attacks (burglary and theft). In the case of the latter, offenders did not present or use any weapons.

The data for silent attacks is quite surprising: more than 80 per cent of the thieves in this study seemed to try their luck despite crew being on board. At the same time, more than 30 per cent of the reported burglars did not care (or know) that crew were on the vessel they intended to break into. One likely reason for this high number of thefts against manned vessels is that cases of petty theft from unmanned yachts were not always reported to the sources used for this study. Nevertheless, this is very relevant data when it comes to making your decision about what to do when you hear something strange at night.

If we interpret the results of this study and use them to draft a rough guideline, skippers would have to expect that in about half of the cases in which their boat is targeted the offenders' goal is to actually confront and control the crew. The other half will merely try to silently snatch items of value from the boat.

If we take a deeper look at the equipment that the offensive criminals (here, the robbers) brought to the scene, the picture becomes more differentiated. See Fig. 1.8.

Types of weapons carried by criminals during an attack

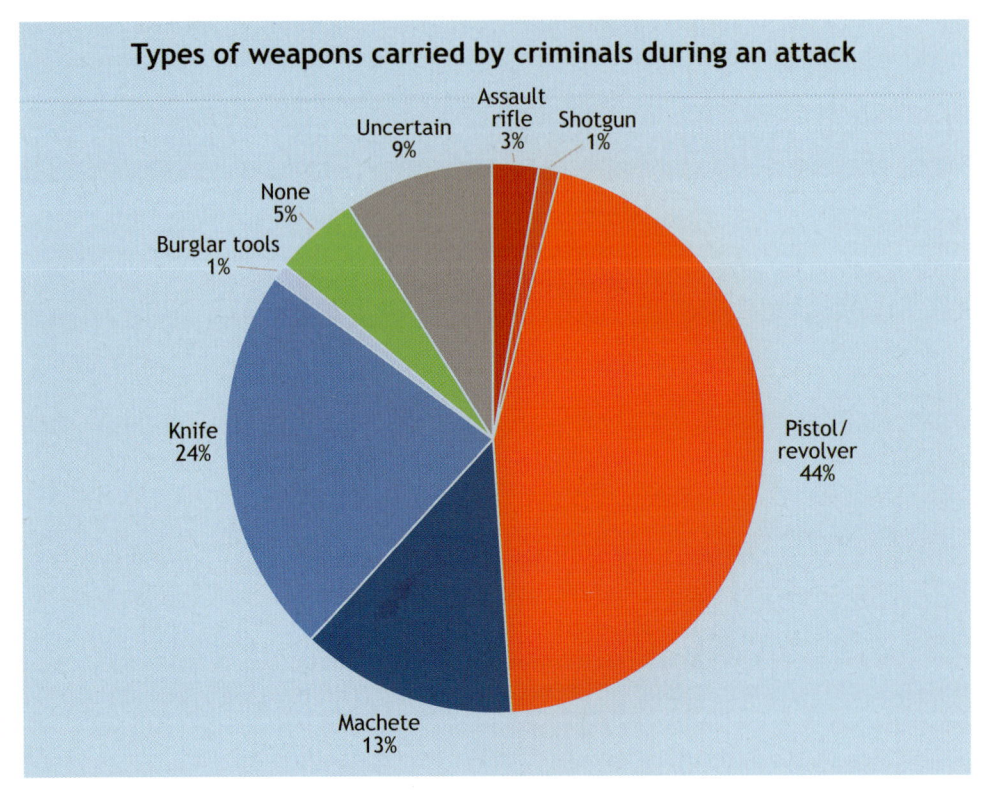

- Assault rifle 3%
- Shotgun 1%
- Uncertain 9%
- None 5%
- Burglar tools 1%
- Knife 24%
- Machete 13%
- Pistol/revolver 44%

FIG 1.8: *Weapons carried by attackers during robberies targeting yachts. (Source: 'Yacht Security 2017')*

Approximately half of the robbers brought firearms with them, making the event extremely dangerous. The other half were armed with machetes, knives, tools or their bare fists, which still made it a dangerous situation, albeit not as unpromising and deadly as those involving firearms.

Keep the big picture in mind: half of the incidents were controlling attacks aimed at confronting and overpowering the crew and half of these were conducted with guns. This means that three-quarters of all nightly visits could potentially be managed by a well-prepared crew without risking a firefight.

The other good news is that most robberies involving guns occurred in very few and very well-known hotspots that can quite easily be avoided by planning your route accordingly.

As of 2017, these regions were:

- Honduras, Guatemala and Panama in Latin America
- St Vincent and the Grenadines and Haiti in the Caribbean
- Venezuela (the worst of them all) and two hotspots in Brazil in South America
- The southern Philippines in Asia.

There are some other areas where criminals with firearms attacked yachts that also have to be watched. However, cases there were few and far between and they are nowhere near as dangerous as the regions listed above.

To summarise, if a yacht is not anchoring or sailing in one of the hotspot regions, the chances of a gun-wielding robber sneaking on their deck are extremely low. Consequently, crews who feel confident and capable do not have to hide below decks until it is all over. A confrontation in which a prepared crew has the initiative can hold more benefits than downsides.

Defenders' tactics: resistance vs submission

How to react when being boarded has long been discussed in a lively manner on the internet and over mellow sundowners around the world. It seems that there are as many strategies as skippers, although some are repeated more often. These include:

The 'Rambo', who storms out whatever the offenders' numbers and weapons.

The 'New York City Smart', who will surrender to gun-wielding robbers but may chase away a burglar.

The 'Submissive', who will always yield to any criminal since he counts on having good insurance that will pay for any damages.

Let's first have a look at some data of past attacks to add some objectivity to the discourse. Two questions will be of interest for defenders of a yacht that is about to or has already been boarded: 'will I be able to fend off the attackers and spoil their success?' and 'will I remain uninjured?'

More-recent works in criminology state that offering resistance to a robber might under certain circumstances very well be a good tactic for getting through a crime situation. Older research that led to the recommendation that you should always surrender to robbers was based just on police records, and didn't take into account the fact that many successful defences had occurred but were simply not reported to law enforcement.

So what about crimes against yachts? When looking at the consequences for a resisting vs a passive crew during robberies, thefts and burglaries the results are quite surprising.

Fig. 1.9 compares how successful crews were at preventing offenders from taking any booty depending on whether or not the crews offered up any resistance. The picture is very clear: resisting crews were quite successful at protecting their property, and the offenders

24 THE COMPLETE YACHT SECURITY HANDBOOK

Resistance and outcome (all attacks)

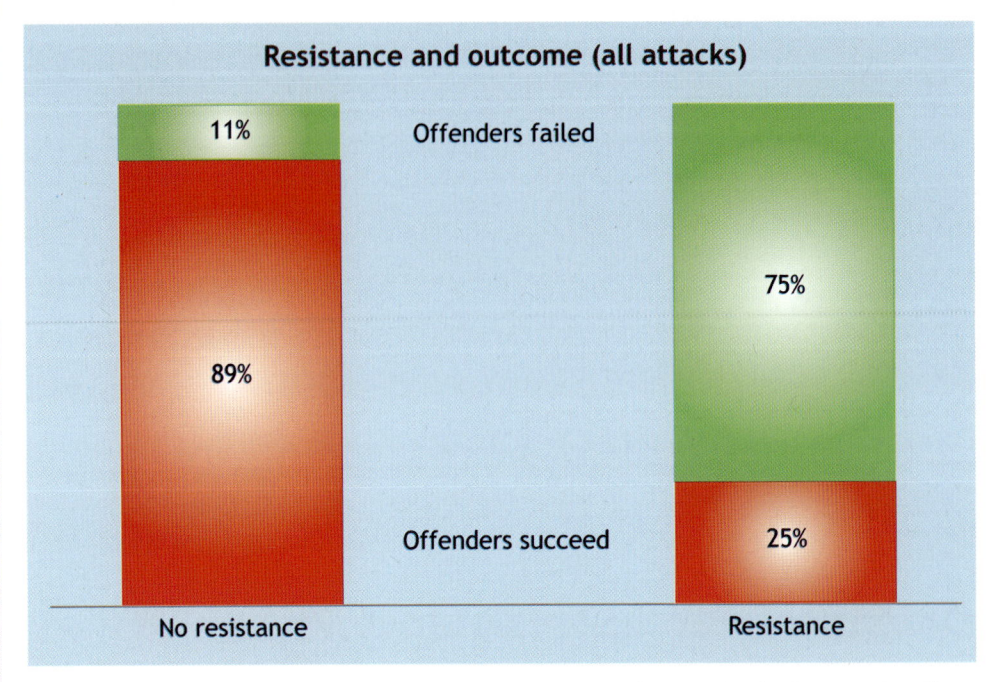

- No resistance: Offenders failed 11%, Offenders succeed 89%
- Resistance: Offenders failed 75%, Offenders succeed 25%

FIG 1.9: Crime outcome depending on defenders' resistance. (Source: 'Yacht Security 2017')

Resistance and injuries (all attacks)

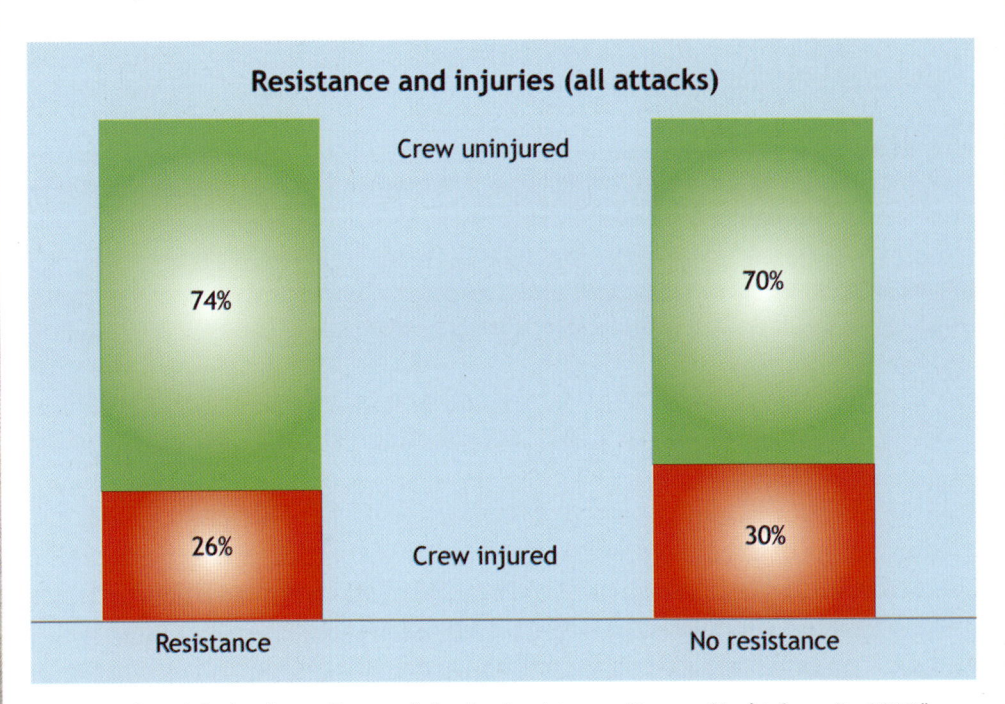

- Resistance: Crew uninjured 74%, Crew injured 26%
- No resistance: Crew uninjured 70%, Crew injured 30%

FIG 1.10: Crew injuries depending on defenders' resistance. (Source: 'Yacht Security 2017')

Resistance and injuries (robberies only)

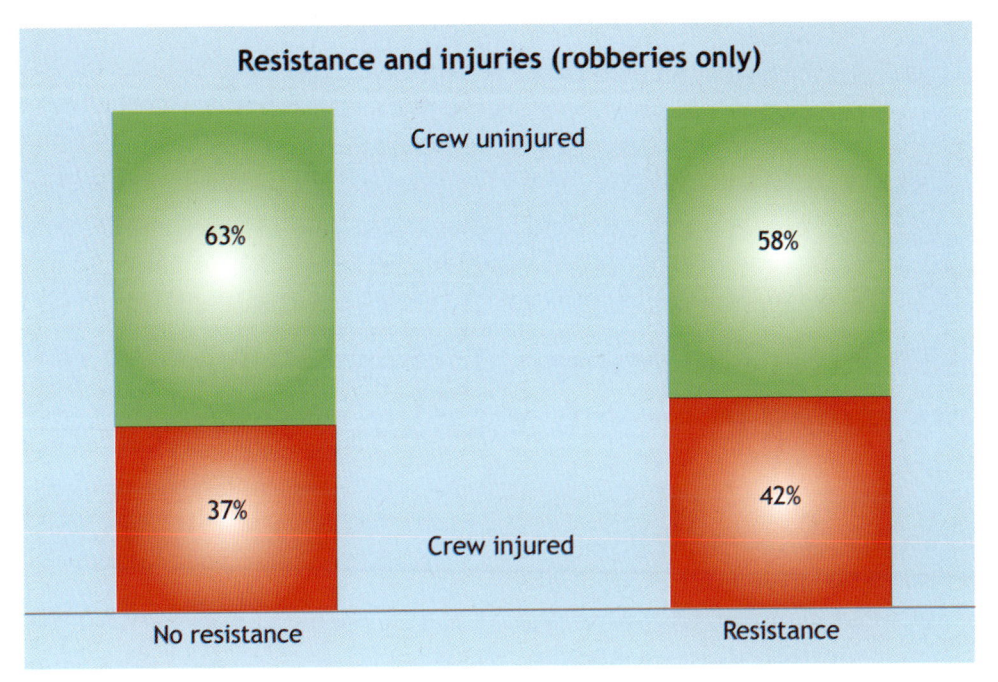

No resistance: Crew uninjured 63% / Crew injured 37%
Resistance: Crew uninjured 58% / Crew injured 42%

FIG 1.11: Crew injuries depending on resistance to robbers. (Source: 'Yacht Security 2017')

failed in their objective in three out of four cases. By contrast, passive crews lost equipment or cash in nine out of ten cases.

The second question that skippers will ask concerns the price that a crew pays when risking resistance. The answer is even more surprising.

The first revelation is the generally low number of injuries that occur during criminal offences. This is due to the fact that we are looking at a combination of theft, burglary and robberies; thieves and burglars do not aim at confrontation and are usually unarmed. The greater surprise is that resisting crews were equally likely to be injured as passive ones. Before trying to explain the reasons for this, let us have a quick look at robberies alone, during which a usually armed assailant boards the yacht to specifically control the crew.

As you can see in Fig. 1.11, the probability that someone will be injured during a robbery is higher than it is when you look at all crime types combined. However, the surprisingly similar casualty figures for resisting and passive crews persists: crews who are passive have essentially the same chance of being injured as a resisting crew. The reason for this can be found by studying the dynamics of a robbery.

Some robbers use violence even before the crew has a chance to voice their submission, simply to make their point. Unfortunately, in many cases the violence did not stop after the crew had surrendered. Instead, the criminals would physically abuse some or all crew members to maintain control and find out about valuables and their hiding places.

The lessons that can be learned from this data lead us to the conclusion that a passive crew will have the same chance of being injured by boarders as a resisting one. However, the resisting crew has an 80 per cent chance of protecting their property, whereas 90 per cent of the passive crews will lose valuables or equipment.

In light of this, some skippers wonder whether there is anything that they can do to improve the odds. Fortunately, there are many strategies and preparations that can be used to reduce the risk of being boarded in the first place (by far the best method). Moreover, when looking at past data there is one tactical aspect that seems to be a true game changer when a confrontation is inevitable: the initiative.

Initiative: The primary ingredient for prevailing in a confrontation

> 'Our strategy was (and still is) to switch on cockpit lights and check before opening the locked door. I failed to follow our procedure. Instead, I opened the door to be confronted and instantly attacked by four criminals, one of them armed with a handgun.'
> Eric, SV *Amarula*, Colombia

Now what is 'the initiative'? A simple definition may sound like this: the one who has the initiative is able to force his actions on the opponent, making him react and thus denying him the opportunity to act himself. Having the initiative usually allows you to decide when, where and by what means you want to confront an opponent.

Three examples:

- Waking up with a knife at your throat is a situation in which you do not have the initiative. The criminal can choose the next action at will.
- Charging screaming towards a criminal who is distracted by a flare your crew has shot through a hatch, armed with pepper spray and a machete is having the initiative. The criminal will have to react to you. Unless he is armed with a gun (and you do not hit him with pepper spray first), he will probably retreat.
- Trying to figure out numbers and armament of boarders walking on the fully lit deck while you are hiding in the dark of your cabin is an undecided situation. You could seize the initiative with a surprising approach from behind, or you could be subdued by criminals aiming their pistols through the windows at you. You will have to take some action to gain the initiative in such a case.

To obtain the initiative, a few factors are very helpful:

- Know the attackers. How many? What is their armament? Where are they?
- Surprise the attacker. Do not let him know where you are, how many you are

and from where you will emerge to engage him. Use distractions such as light or noise.
- Overwhelm his decision-making process. Start loud noises such as the foghorn, trigger the DSC distress call on your VHF and deploy flares and/or strobe lights prior to the actual confrontation. Such elements distract and make him think about possible support arriving from the shore or other boats.
- Be stronger or better armed. The more menacing you are or appear, the more you will be able to dictate the course of action. However, this is not an imperative. Poorly armed single-handed skippers have successfully chased away a group of boarders, and crews have managed to fend off gun-wielding criminals with bare hands when using the element of surprise.

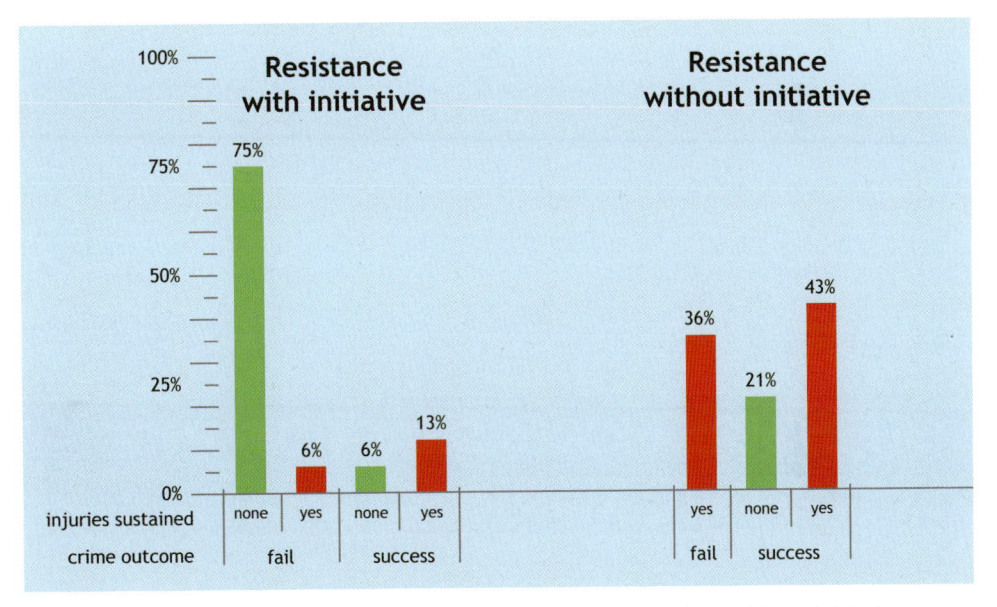

FIG 1.12: Crime outcome and injuries depending on resistance with or without initiative. (Source: 'Yacht Security 2017')

Initiative is not something that sticks with you once you have obtained it; rather, you need to defend it actively until the skirmish is over. The best way to keep the initiative once the actual confrontation has started is through continuous offensive action against the opponent until the threat has been neutralised. In the case of yachting, this usually means that the offender has gone overboard, where he cannot do much harm for the moment.

Offensive action means a combination of speed and aggression. Be loud and violent until the attackers yield or, better yet, leave. Yell at the top of your lungs. Be as well armed as your equipment allows. If you only have non-lethal weapons, use them right away and keep on using them until the criminals have been subdued or are in the water. The use of lethal weapons involves various legal and ethical factors and thus it is not recommended without differentiation at this point.

Let us take a look at how the scene changes when resisting crews have the initiative (or not).

The group of columns on the left shows the outcome of robberies where crews resisted with initiative. By adding the two numbers in the subgroup "fail", we can tell that these crews were able to prevent the robbers from grabbing booty in 81 per cent of the reported cases. By combining the two green columns in the group "resistance with initiative", we conclude, that 81 per cent of the crews were not harmed during the confrontation.

By contrast, crews trying to resist from a position without initiative did not fare very well. The group of columns on the right of this chart shows the results: They were only successfully repelling the criminals in 36 per cent of cases. At the same time, the risk of being injured when resisting without initiative rose to a tremendous 79 per cent. Put differently, confronting criminals without having the initiative was very dangerous: the risk of being injured was very high and the chances of successfully repelling the criminals were quite low.

The crew from SV *Amarula* had great precautions defined to help them keep the initiative: a locked companionway would deny boarders to surprise the crew off-guard and having cockpit lights on during the night would expose attackers to the crew and potential supporters from ashore. By – very humanly – abandoning their plans in the heat of the moment, they lost initiative the moment they opened the companionway door unprepared.

OFFENDERS, MODES OF ACTION AND REACTION TO RESISTANCE

Cruising the global oceans, you may encounter offenders who differ as much as people and cultures do. Origin, early years, criminal history, ability to plan, behaviour when confronted and the general inclination towards violence are discriminating factors.

Ultimately, the only thing that is of relevance to skipper and crew is the way in which an offender is 'functioning'. Crews will need to know where they might encounter what type of attacker and how to deal with him once they confront him.

There are a couple of categories that may prove useful when trying to assess who you are dealing with and how to plan, since you will meet some types only in well-defined regions.

Keep in mind, however, that regional peculiarities have a modifying impact on offender types. A prominent example of this idea is the general availability of firearms in some regions – a factor that can convert the simplest ruffian into a murderous opponent. Regions that have suffered from war and widespread terrorism usually produce more violent criminals than relatively peaceful countries.

FIG 1.13: *Organised piracy is effectively countered in regions such as Somalia and the Strait of Malacca. (Photo: Getty)*

The 'A clear picture on risk: cruising area and yacht situation' section (see pages 39–45) describes these peculiarities and ways to discover them in detail.

Let's have a look at the most relevant offender categories.

The 'Seduced'

This category includes all offenders who do not have any prior criminal history and who are usually not part of a criminal environment. The seduced do not aspire to a criminal career. They are most likely well integrated in their local social networks and do not have access to criminal contacts who could serve as channels to monetarise any booty acquired during their crime. Typically, these individuals will be adolescents or young adults, mostly male.

An attractive item to which there is easy access may trigger the crime. Typical objects include an unlocked bike on the dock, a longboard or a case of beer stored on deck at arm's length from the pier, or a jerrycan unsecured at the lifelines.

In some cases, a trial of courage set by a questionable peer group may be the motivation for the crime. Alcohol or other 'soft drugs' may also have helped to reduce any inhibitions on the offender's side.

This type of offender is present in most regions of the world and he will be one of the most likely attackers in safe regions (Risk Level I).

In the West, the seduced are usually encountered in marinas. Wherever life at or on the water is widespread and the normal state of being (e.g. island states such as Indonesia, coastal regions with stilt villages, etc), the seduced may also try their luck with yachts at anchor.

An offender from this category will only rarely board a yacht; instead, he will typically try to grab any booty from dry land or from his own vessel.

Consequently, the seduced are mostly thieves. They commit their crime alone, rather than in a group. A weapon is not to be expected unless carrying arms is a regular habit in his society.

It is extremely unlikely that the seduced will fight when confronted; instead, they will abort their actions and try to escape. This means that they can usually be chased away by a determined and robust defence. An exception to this general behaviour could be met should an attacker be under the influence of stronger stimulants, such as cocaine or methamphetamine.

The inexperienced

Inexperienced criminals are individuals who have taken their first steps in a criminal career. In contrast to the seduced, they approach the vessel explicitly to commit a crime. They are more likely to carry tools to overcome entry barriers or devices to secure your valuables on deck. However, they are not yet professionals and thus they do not fully understand the consequences of a discovery. There is a high probability that they have not planned escape routes or confrontations and lack experience in overpowering a crew. Inexperienced criminals may be familiar with general violence in their social environment and thus could display fewer scruples in a confrontation than the typical yacht crew.

This group aims at booty both for their own use and to sell to a network of receivers of stolen goods. Their target items will be electronics, yacht equipment and cash. The inexperienced will most likely be younger males.

Lacking a mature criminal network and the experience to plan and coordinate combined attacks, they often work alone. There is a moderate possibility that they are drug addicts and commit the crime to finance their addiction. If they are under the influence of a hard drug, their behaviour will be difficult to predict (see page 33, 'The disoriented and offenders under the influence of drugs').

You will meet this type of criminal in all regions. They are most likely thieves and burglars, rarely robbers.

If you discover a lone inexperienced criminal early and without having lost your initiative, determined opposition will result in their escape.

In regions with a wide availability of firearms, the inexperienced thief or burglar is more likely to compensate for his lack of routine with the use of guns for 'self-defence' or, in the case of an intended robbery, as a method to control the crew. The use of firearms makes the offender much more dangerous.

Organised crime and professional pirates

This category includes experienced and sometimes specialised criminals. It ranges from professional robbers and burglars to the type of pirates who roam the waters of West Africa and Malacca. Professional criminals take major risks, although they are more coordinated and better prepared than the offenders introduced thus far. They will most likely try to scout their targets before the attack in order to gain a better understanding of what benefits and risks are involved in an intended action. However, like their less-practiced colleagues, professionals will also make an impromptu strike if an easy opportunity is presented.

Pirates are experienced in all aspects of the criminal offence that they commit, including the use of violence to control their targets. Consequently, they are likely to employ threats and actual violence to achieve their goals.

Professionals are tightly set in a criminal environment and can rely on a network of receivers to whom they can sell their booty. They may be acting on orders to strike a specific yacht or item. Some groups have a sophisticated infrastructure that will allow them to monetarise yachts and imprison a band of hostages for a longer time.

Professional criminals will most likely commit burglaries and robberies with at least one or more accomplices. The size of the criminal group attacking a boat can thus provide information about the grade of experience that the attackers have and how they will react to resistance.

Especially in higher-risk regions, professionals will be robbers, whereas in safer regions they may strike as thieves and burglars.

As for other types of criminals, conspicuous and visibly effective entrance barriers combined with alarm systems can help to prevent an attack by these offenders. The same holds true for an early detection and determined resistance from a position of initiative.

As soon as a group of professionals is on board, repelling them will be very difficult. It is likely that they will be much more experienced and less scrupulous in overpowering others than the typical yacht crew. That said, if the crew can maintain the initiative and inflict elements of stress on the offenders (such as howling sirens and

FIG 1.14: Raiding a suspected pirate boat. (Photo: Creative Commons)

bright lights), they may stand a chance. Whatever the tactics used by the crew, however, repelling professionals that are already on board carries high risks.

Terrorists and combatants

Terrorists and combatants in civil or regular wars are a special group of attackers. They differ in terms of their behaviour and their experience, armament and equipment. Their motivation is also different from that which drives your typical criminal.

Although terrorists can be motivated by greed for money, their primary goal is media coverage and drawing attention to their political or religious goals. Abduction, murder, gruesome mayhem or spectacular destruction serve these goals very well. This poses a special situation for yacht crews, since terrorists may not aim at boarding and overpowering at all; they could simply destroy the boat and kill all crew from a distance so that they can send footage of the deed to international media networks. Fortunately, such types of attack have not yet occurred against yacht crews.

Instead, well-organised abductions to gain maximum amounts of ransom from Western companies, governments or the victims' families were the preferred method between 2011 and 2017. Depending on the availability and tactics of local rescue operators as well as your home country's strategy concerning negotiations with terrorists, the abduction could not only prove hellish but also fatal for skipper and crew. For those interested in researching the plight of terrorist hostages, some good books written by former victims have been published in recent years (see the Reference section on page 188 for a selection).

FIG 1.15: *A group of Abu Sayyaf terrorists in the Philippines – Muamar Askali (second from right) was killed in 2017.*

Parties of a civil war also sometimes resort to terrorist tactics. These actions mostly target the civilian population of the adversary rather than third parties, such as visiting yachts.

Both terrorists and combatants are usually very highly trained in military tactics and methods. They will be well armed with assault rifles, explosives and the like. Do not expect any scruples when being approached by terrorists: it will be a life, life as hostage, or death situation and it is extremely dangerous to repel them. A very tragic example of what can happen can be seen in the attempt by the crew of the German-flagged yacht *Rockall* to repel just such an attack in 2016, which resulted in one crew member being shot and killed while – as stated by one of the terrorists – attempting to confront the attackers with a gun. The other was abducted and held hostage to be killed when the ransom wasn't paid.

The only realistic hope for a regular yacht crew is early intervention by a determined military or police force. Crews should therefore focus their response to an impending attack on delaying the boarding as well as raising help from any military or police units that may be nearby.

Normally, this group of attackers does not show up out of the blue in any given region. Sustained terrorist activity and wars (both international and civil) are usually covered in the local and international media and thus can easily be researched by skippers and crews. Unfortunately, though, sometimes crews become the first victim of a new terrorist threat or a surprising change in an existing group's tactics. For instance, in 2014, Abu Sayyaf terrorists (a Philippine group known for abducting Westerners for ransom) surprised the unfortunate skipper Stefan Okonek and his companion Henrike Dielen in Palawan, 480 kilometres (300 miles) away from the group's base of operations. Up to this point, no one would have expected such long-distance strikes. Since then, the safety circles around Jolo and the south of Mindanao have been extended and will remain that way until the threat has been contained by effective actions against the group. Regions that are troubled with terrorist activities and war are classified as the highest risk level in this book. It is explicitly recommended that you stay away from such regions.

The disoriented and offenders under the influence of drugs

Disoriented offenders are sometimes under the influence of drugs or are mentally ill, and generally commit thefts, burglaries, vandalism and robberies. Both sub-types are known to attempt attacks involving risks that are irrational, bordering on stupid, when judged from the perspective of a healthy crew. Some reports describe attempted theft in broad daylight with the crew aboard. Others speak of individuals who watched a boat anchored in 'their bay' for some time then simply picked up an improvised weapon, jumped in their vessel and tried to rob the yacht.

Many cases of attacks that are committed by disoriented offenders are reported in higher-risk regions where drugs are more common and healthcare systems are weak, although offenders currently under the influence of drugs can be encountered in all regions. In the West or generally safer waters, they might prowl marinas and look for easy-to-grab items. It is highly unlikely that they will attack anchored boats given that they are usually incapable of handling a boat or swimming longer distances.

The first contact with a drugged offender can be fierce on his side, although usually they retreat quickly when they meet determined resistance. An exception to this rule is attackers who have consumed stimulants such as cocaine, methamphetamine or khat to specifically prepare for an attack. This type of drug-user will look less ragged than your typical disoriented offender or those who attack to finance a heroin addiction. They will also be more aggressive and take a longer time to retreat even when confronted with fierce opposition.

Some of these attackers are tempted by the wealth of a yacht, whereas 'foreigners' in their bay provoke others. Whatever the motivation, due to the person's general mental state the attack will most likely be committed in a clumsy manner, involving awkward behaviour and communication. They almost always act alone and more or less spontaneously. If they are not stopped during their attack, they are capable of causing extensive damage and might even proceed to injure a crew that is inactive or compliant. It is therefore important to prevent them from boarding at all costs. Once on board and confronted, they will most likely not react rationally and crews lacking training in psychology will find it difficult to calm them down. Moreover, physical attacks against objects and people can spontaneously occur, even if the offender seems to have already calmed down.

In order to avoid this situation, drugged or disoriented offenders can be repelled if they are spotted early and met with a strong show of force, using clear words, loud yelling and – if everything else fails – presentation of weapons to support your determination.

In instances when you and your supporters on land do not succeed in containing the threat (most likely by a police arrest), the disoriented attacker will still pose a threat as long as you are present in the area. If there is no chance to have him removed from the scene, moving your boat elsewhere might be the better alternative.

PHYSICAL CONFLICT AND SELF-DEFENCE

This book focuses on avoiding confrontation with criminals by means of route planning, preparation and defensive strategies while travelling. Nevertheless, it also covers defensive weapons and encourages readers and skippers to consider resistance in some situations.

Readers have to keep in mind that the use of weapons and physical violence bears risks for their own health and life as well as those of the people they engage. None of the tactics described here should be used in any situation other than self-defence. This will keep both your conscience and your criminal record clear. Remember: self-defence, as defined in most countries, is the use of **reasonable measures** to end a **dangerous** and **ongoing threat** against oneself, other people or sometimes valuable assets.

Be sure that you are truly threatened before you act, and then only act within reasonable limits, especially if you carry dangerous or even lethal weapons. Stop immediately when the threat is over. Leave the area instead of trying to pursue, arrest or even 'punish' an attacker. Do not get carried away: injuring, let alone killing, fellow humans by accident or outside of the limits of self-defence is in no way advocated by this book.

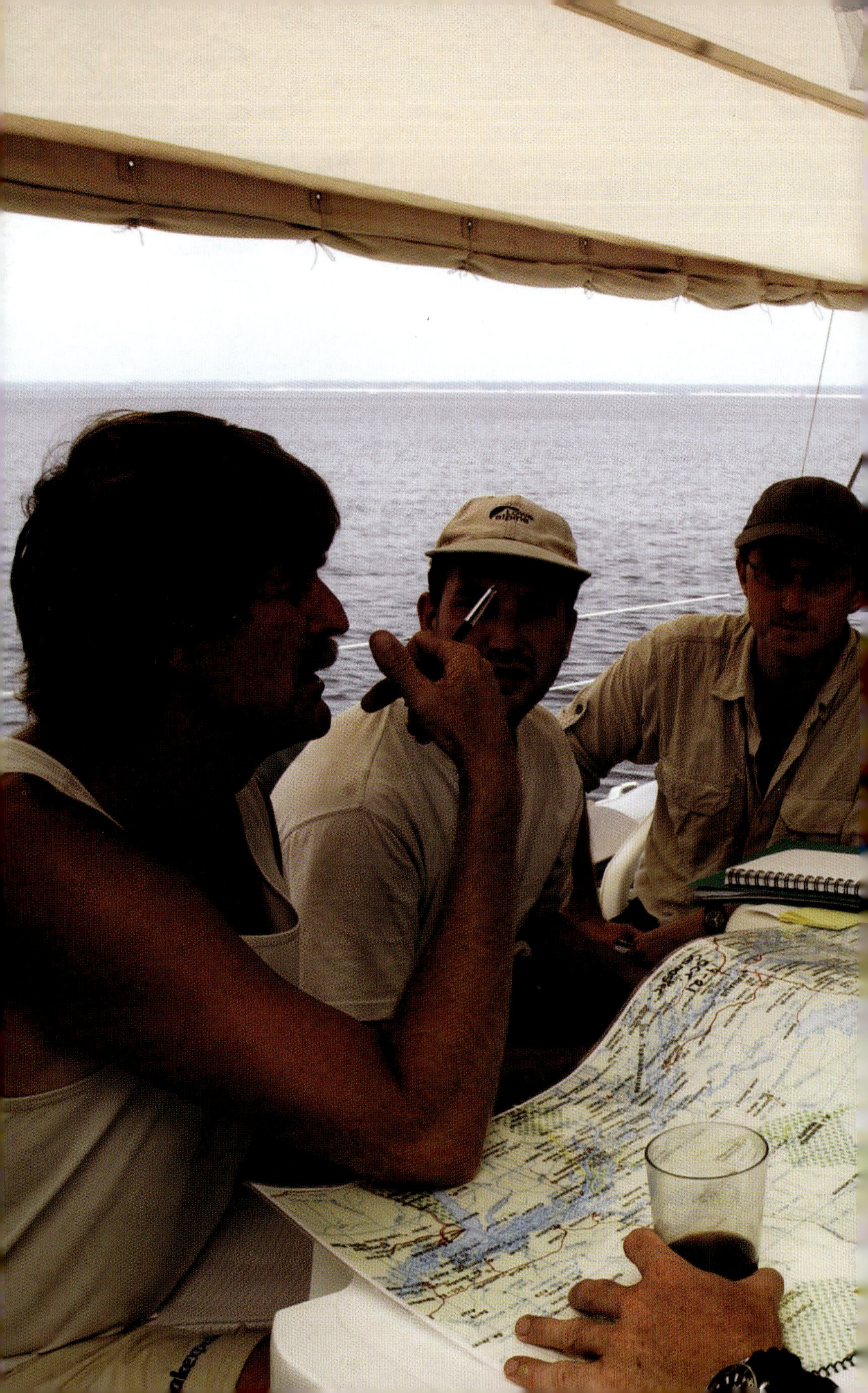

Part II

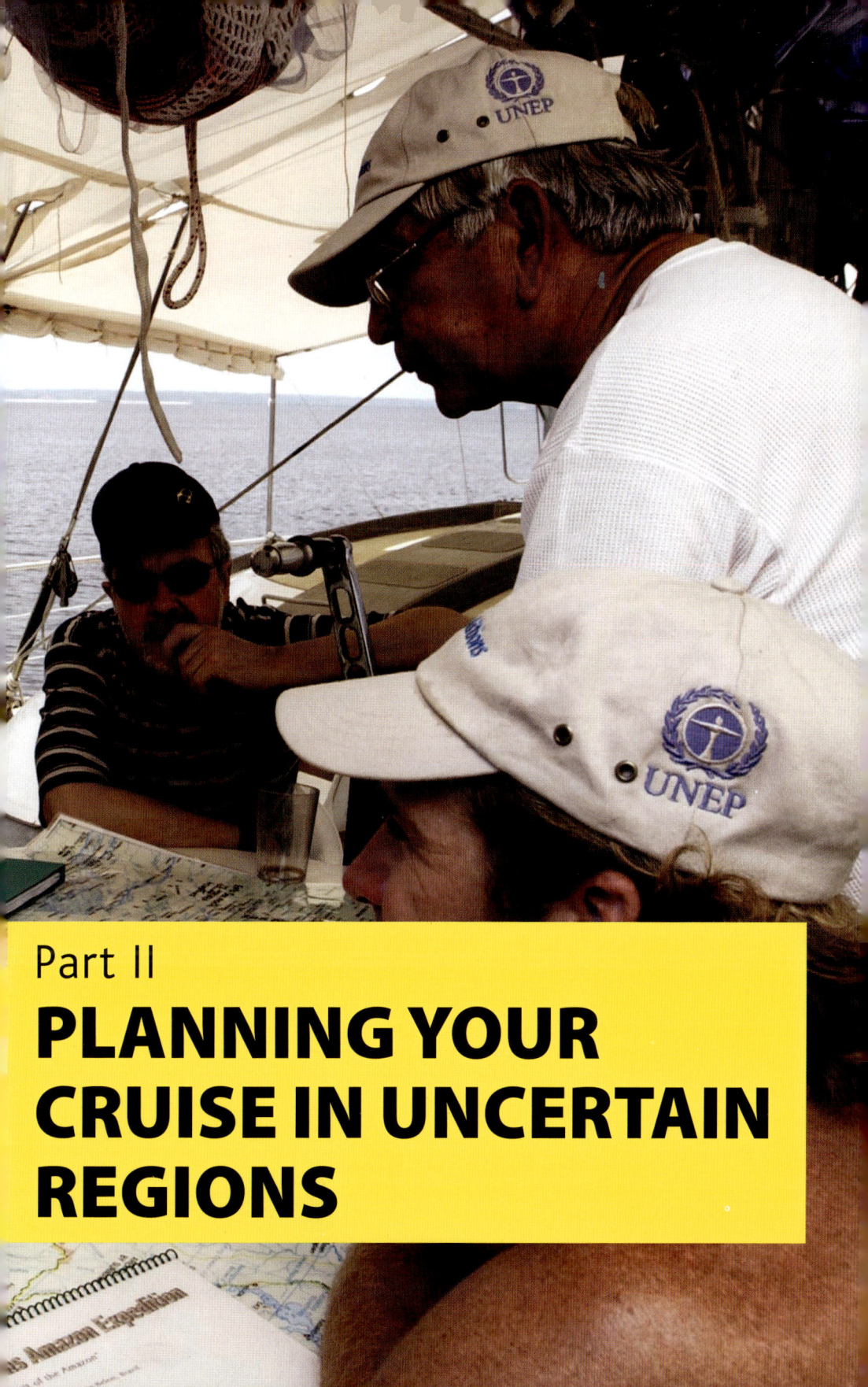

PLANNING YOUR CRUISE IN UNCERTAIN REGIONS

> The Security Balance is a simple tool to help skippers asses as well as balance risks and countermeasures.
>
> Quick and simple methods can be used to help in the understanding of the factual risks of a region and demonstrate how they can be countered by adapting crew, equipment and vessel type. If balancing these proves unfeasible, a change of route and destination will most likely be the best strategy to enjoy a carefree voyage.

Good planning is by far the best method to ensure relaxed and safe sailing. A proper plan will ensure that you either restrict your voyages to regions where your capabilities at least match the risks or avoid any serious trouble altogether.

Many long-term cruisers and future liveaboards wonder about how to prepare boat and crew for a cruise to unknown and potentially dangerous waters. The more responsible charter skipper planning a trip to the Caribbean might also ponder similar topics before taking friends and family aboard.

A simple tool – the security balance (see Fig. 2.1) – provides a quick-and-easy way to establish how a ship and crew should be prepared according to the intended cruising ground and the general situation a yacht be in.

SECURITY BALANCE: A SIMPLE METHOD FOR ASSESSMENT AND PLANNING

You can think of the security balance as a set of scales. One tray holds your intended cruising ground and the yacht situation, defining the risks of the trip. Crew capabilities, yacht type and her equipment are piled on the other side, acting as security measures and counterweight.

The more crime that you expect in the area that you would like to visit, and the more time that you spend at anchor rather than in safer marinas, the more weight will be on the risk side of your scales. You can counterbalance this by choosing a sturdier yacht, equipping her with adequate security equipment and staffing her with a more experienced, capable crew. Whenever the risks are not properly countered by security measures, skippers will most likely feel a little more uneasy during the nights in solitary anchorages. They are also at a higher risk of falling victim to a successful thief, burglar or robber.

By using the security balance, even less-travelled sailors are able to assess and optimise their security situation in just a few steps.

PLANNING YOUR CRUISE 39

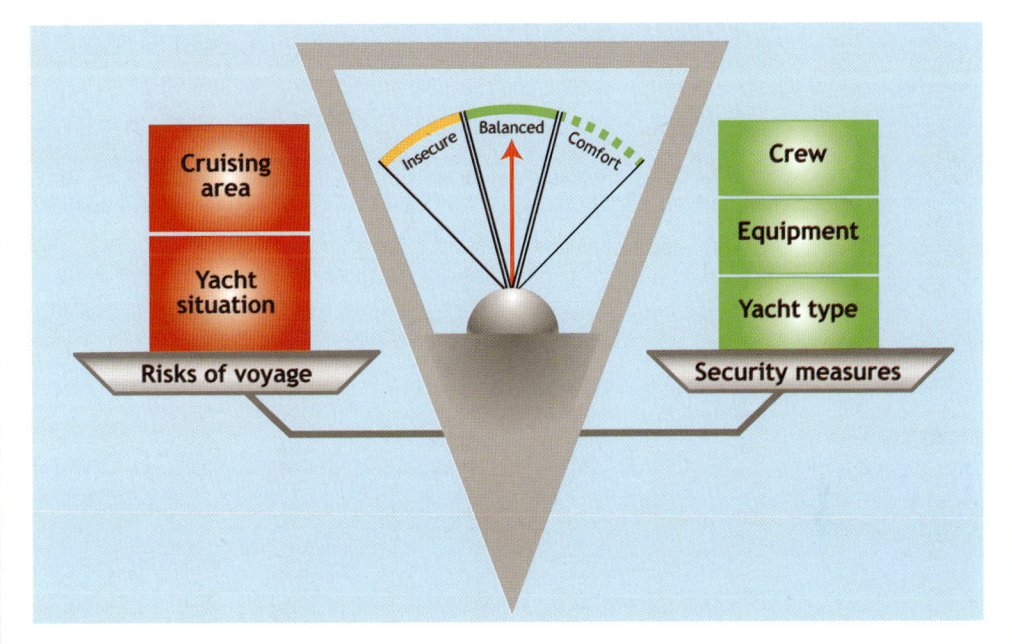

FIG 2.1: *The security balance weighs risks of regions and yacht situation against security measures.*

A CLEAR PICTURE ON RISK: CRUISING AREA AND YACHT SITUATION

The intended cruising area is the most important aspect for safety on a cruise. The spectrum spreads from mostly safe coasts in the northern hemisphere to pirate-plagued waters such as the Gulf of Guinea or the sea around Basilan, the Philippines. When travelling the waters of northern Europe, northern America or Japan, security on board is mostly as good as it ashore. With a bit of reasoning, sailors will experience few relevant security risks. However, things change when sailing to the Mediterranean Sea or the Caribbean: petty theft in the marina or break-and-entries at anchor are commonplace in these areas. Some islands of the Caribbean are now the sites of increasing violent crime, while some countries on its fringes – such as Venezuela – are currently plagued by extreme poverty and reckless gunmen.

For ease of use, this book categorises the cruising areas into four risk levels. These allow skippers to gain a quick idea about the types of crimes to be expected as well as the probability of a certain crime type occurring.

The higher the overall risk of pleasure crafts being targeted by criminals, the higher the risk level in this classification.

	Marina			At anchor		
Risk level I	Burglary	Abduction		Burglary	Abduction	
	Theft	Robbery	Terrorism	Theft	Robbery	Terrorism
Risk level II	Burglary	Abduction		Burglary	Abduction	
	Theft	Robbery	Terrorism	Theft	Robbery	Terrorism
Risk level III	Burglary	Abduction		Burglary	Abduction	
	Theft	Robbery	Terrorism	Theft	Robbery	Terrorism
Risk level IV	Burglary	Abduction		Burglary	Abduction	
	Theft	Robbery	Terrorism	Theft	Robbery	Terrorism

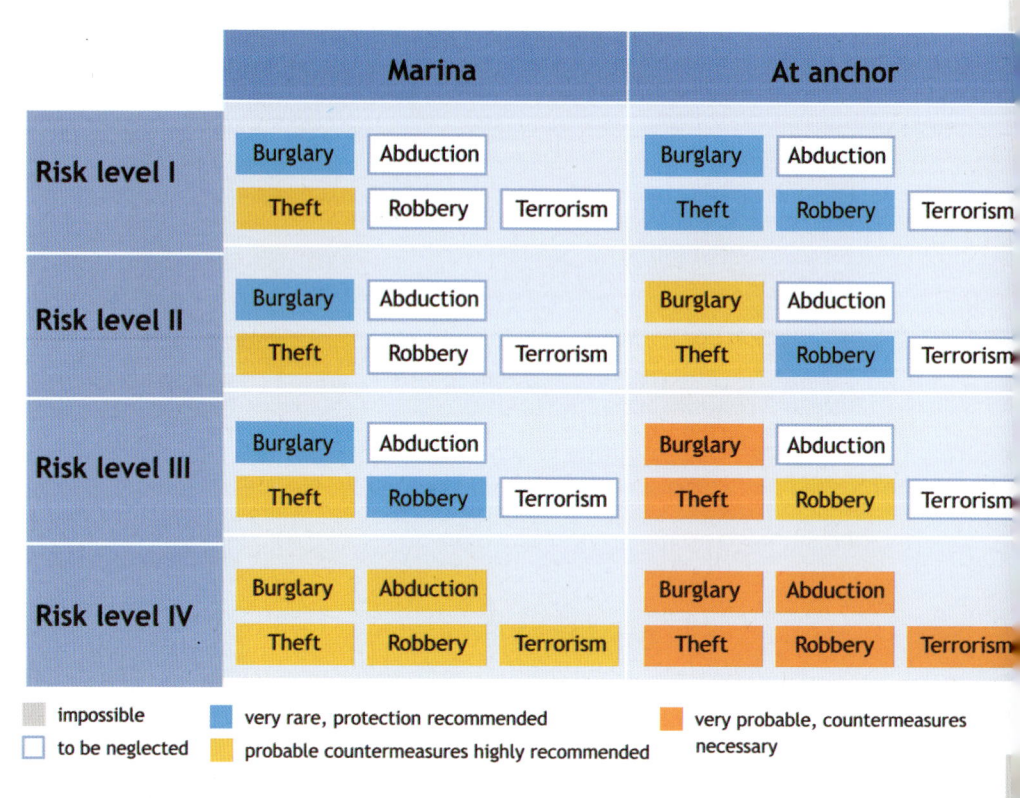

- impossible
- to be neglected
- very rare, protection recommended
- probable countermeasures highly recommended
- very probable, countermeasures necessary

FIG 2.2: *Crime types and probability according to a cruising area's risk level and yacht situation.*

As you can see, the table in Fig. 2.2 already includes the second basic risk factor of your cruise, namely the yacht situation. This is essentially the type of place in which you spend most of your time while cruising.

The colour codes provide information on how probable a criminal attack on a pleasure yacht will be. The judgement is based on the occurrence of attacks in the past as well as from current indicators for crime (i.e. economic status, general proliferation of violent crimes in the country, etc).

Keep in mind that even if the general probability that some crime may be committed is high (and you should therefore do some preparation), in some places this does not mean that a region is off limits or totally spoiled. Unless a cruising area is categorised as Risk Level IV, a well-prepared crew on a well-equipped boat should feel confident to visit.

So let's have a look at the current allocation of larger cruising regions to risk levels.

Regions and countries are very diverse. Consequently, some of the countries with a grim reputation or an overall high risk level feature marvellous save havens (such as Île-à-Vache in Haiti at the time of writing) and many mostly safe countries offer a few areas that are notorious for armed crime against yachts (such as just Itaparica in Brazil in the recent past). These regional or local effects can only be discovered if skippers delve a little deeper.

At sea

Burglary	Abduction	
Theft	Robbery	Terrorism

Burglary	Abduction	
Theft	Robbery	Terrorism

Burglary	**Abduction**	
Theft	**Robbery**	Terrorism

Burglary	**Abduction**	
Theft	**Robbery**	**Terrorism**

How to interpret the Risk Level Chart:

If you plan to visit a Risk Level III area, you will be mostly safe from serious crime while in the marinas. Keep a clear deck, as theft is not uncommon.

If you plan to anchor, you will have a higher risk of burglary, theft or robbery. Use crime prevention measures. Prepare ship and crew for possible nightly boarding.

Attacks while at sea are very rare, but have occurred in the past.

Some may be surprised to find that very poor societies such as Vanuatu have the same risk level as the Netherlands, or Thailand is in the same category as the southern coast of France. One of the most important factors that contributes to the development of sustained crime is a large inequality between the rich and the poor in a country. However, cultures cope differently with such inequalities. Take some of the societies found in the Polynesian regions of the Pacific: while most inhabitants would be considered very poor by Western standards, (very sensibly) most local Polynesians do not actually care a bit about these standards – they love their way of life and simply do not strive to be what Westerners would call wealthy. In other cultures, it is more readily accepted that you can obtain wealth by taking it from others. In such regions, the inequalities do not have to be substantial to trigger criminal behaviour.

Compare Thailand and the Dominican Republic, for instance: both countries are very similar in their general economic status (as measured in GNP per capita in 2016), yet the chances of an assault on a yacht are substantially higher in the Dominican Republic. If you compare Fiji and the economically similar Honduras, you will notice that there are few crimes against yachts in Fiji whereas skippers have to be cautious in the waters around Honduras. Without attempting to theorise on the reasons, this data seems to indicate a higher tendency for crime in Latin American societies compared with some other countries around the world that perform equally well in economic terms.

Risk Level IV regions are special: the risks are so high that the usual pleasure craft crew with a standard boat cannot counterbalance them. Only the presence on a ship of armed and trained professionals with at least some experience combating determined military assaults would put sufficient weight on the scales to equalise such risks.

Risk level I	Coasts of Northern Europe (North Sea, Baltic Sea, northern Atlantic to Spanish coast, England, Iceland and Greenland) Coasts of the USA and Canada Australia, New Zealand, Pacific from Galapagos to French Polynesia Singapore, Japan and South Korea
Risk level II	Coasts of the northern Mediterranean Bora Bora Morocco, Tunisia Northern Caribbean (BVI, USVI), Martinique, Guadeloupe Thailand and Malaysia, China Most of Mexico, Chile and Peru
Risk level III	Rest of Caribbean including Great Antilles and Haiti (Île-à-Vache) Solomon Islands, Papua New-Guinea, Indonesia Coasts of Africa (excluding northern Indian Ocean and Western Africa north of Namibia) Most parts of northern and eastern coasts of South America Latin America Red Sea, Socotra, Israel Rest of Indian Ocean Special vigilance in Bangladesh, Honduras, Guatemala, passage from Trinidad/Tobago towards South America, St. Vincent and the Grenadines
Risk level IV	Southern Philippines (350-mile radius around Jolo/Basilan) Coasts of Somalia, Nigeria and bordering nations (Gulf of Guinea) Coast of Venezuela Pakistan, Syria, Yemen (excluding Socotra) Haiti (excluding Île-à-Vache) Libya

FIG 2.3: *Rough allocation of cruising areas to risk levels as assessed in 2018.*

When researching which countries to visit in the planning stages, you will find countries in Risk Level IV that are in a state of war, civil war and civil unrest, where there is active terrorism, and that are home to societies that feature radical hostility against Western values. Entering such waters on a yacht – especially bikini clad or wearing your favourite college-team shirt – would be very daring indeed. Due to the high risks as well as the mobility of terrorists, war parties and organised crime gangs in these zones, we do not encourage attempting to visit even apparently safe sub-regions of Level IV areas. Try to give them a wide berth. If this is impossible, consider all the measures outlined in the 'Sailing in pirate-plagued waters' section (see page 130). Apply the same caution when sailing the 'High-vigilance Zones' marked in Fig. 2.4.

Finding more-detailed and up-to-date information that you can use to inform your judgement is sensible, and requires just a few hours of your time for each area in question. Most of the research can be done on the internet. Look for the number of crimes and the quality (theft, burglary, assault and capital crimes) of crime against yachts and use this data to update or refine the classification of your cruising ground yourself.

The primary sources of information are websites of state departments or the offices of foreign affairs. You may want to start with your own country's systems for a first impression, but it is also useful to check out several sources so you can compare and contrast the

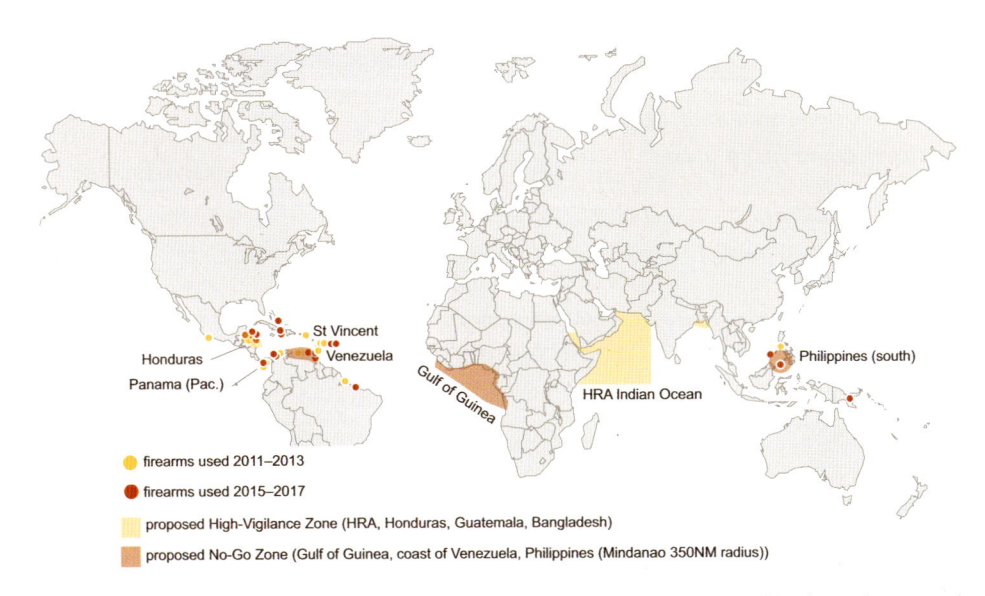

FIG 2.4: *Use of firearms during assaults on yachts (2011–17) and proposed high-vigilance and no-go areas. (Source: 'Yacht Security 2017')*

findings and start to build up a detailed picture. The information issued by the British government has proven precise and quite specific in the past. This is because rather than limiting the information to a country level, the British issue region-specific warnings and, in some cases, information especially for seafarers. If you combine the data found on two to three sources, you will most likely gain a good, albeit conservative, first impression.

To delve a little deeper and obtain a clearer picture that is more relevant for pleasure crafts, check out www.noonsite.com. This site is aimed at the global cruising community and features rich and often recent information for sailors. Many sailing magazines nowadays provide their archives online, which means that you can sometimes find more-or-less recent information on cruising areas on their websites. It is helpful to check out touristy titles as well as the 'hard-boiled' blue-water sailing magazines since both may have featured a piece on your destination in the recent past.

Another great source for your research is the blogs of other sailors. Check Google for keywords such as 'cruising, xxxx [insert name of your destination], blog'. Many skippers report invaluable information, providing tips and great security intelligence for even the remotest places. The chances of you finding information about specific bays and anchorages using these blogs are often also good.

Quickly scanning the local press websites of your intended cruising area is usually a valuable add-on. Especially if you have discovered some incidents in other sources, you may want to follow up in the online version of local papers. Here, you can gain a good feeling for how the community reacts to crime, whether the police seem to do their best to reduce

> **AVOID AREAS WITH GUNS**
> Avoid any region where criminals have used firearms during attacks on tourists and yachts in the recent past!

incidents and if the assailants have actually been caught. Reports of petty crimes in anchorages that might not have made it to the international community may be discovered here, too.

Finally, you may wish to check the available cruising and pilot guides for your cruising area. Depending on the target region, these can be quite reliable. Some – such as Chris Doyle's guides to the windward and leeward islands in the Caribbean – are issued every other year and are usually well informed on security issues. High-quality sources for remoter regions such as the Western Pacific and parts of Asia are harder to come by since most available guides either do not cover security issues or are simply outdated.

Apart from checking out general reports concerning yacht security, it is also important to develop a fine sense for crimes committed using firearms. Whenever you discover a place where such assaults have occurred within the last 12 months, try to avoid it.

It is almost impossible to defend against a gun-wielding attacker. Even if your crew carries firearms, any conflict with such an assailant would be extremely dangerous. This strict rule might be relaxed only if you can positively discover news that the assailants have been arrested and taken off the scene for good.

Having done this research – starting with general information and ending with local or recent cruisers' reports on specific areas – you will now have a clearer sense of the crime quality and the probability of it occurring. Let us now take a look at the second-most important factor for security, namely the yacht situation.

In the Risk Level Chart above (see Fig. 2.2) the major yacht situations have already been introduced: marina, at anchor and at sea. These situations include some subsets that are not very important for planning at this point but might be kept in mind when you actually visit a place.

- Marinas in this book include public docks, if they welcome pleasure yachts. They do not include industrial or dedicated fishing ports.
- The category 'at anchor' also includes ships that are tied to public mooring balls that are not part of a marina (and consequently not part of their security scheme), since the two situations pose similar risks – weak security, solitude, no direct access to land-based emergency infrastructure, etc.
- The category 'at sea' comprises situations en route. This includes voyages in coastal as well as international waters.

Our recent study analysing more than 250 criminal attacks on yachts worldwide provides a good overview of the general distribution of crimes against ships by situation – see Fig.

PLANNING YOUR CRUISE 45

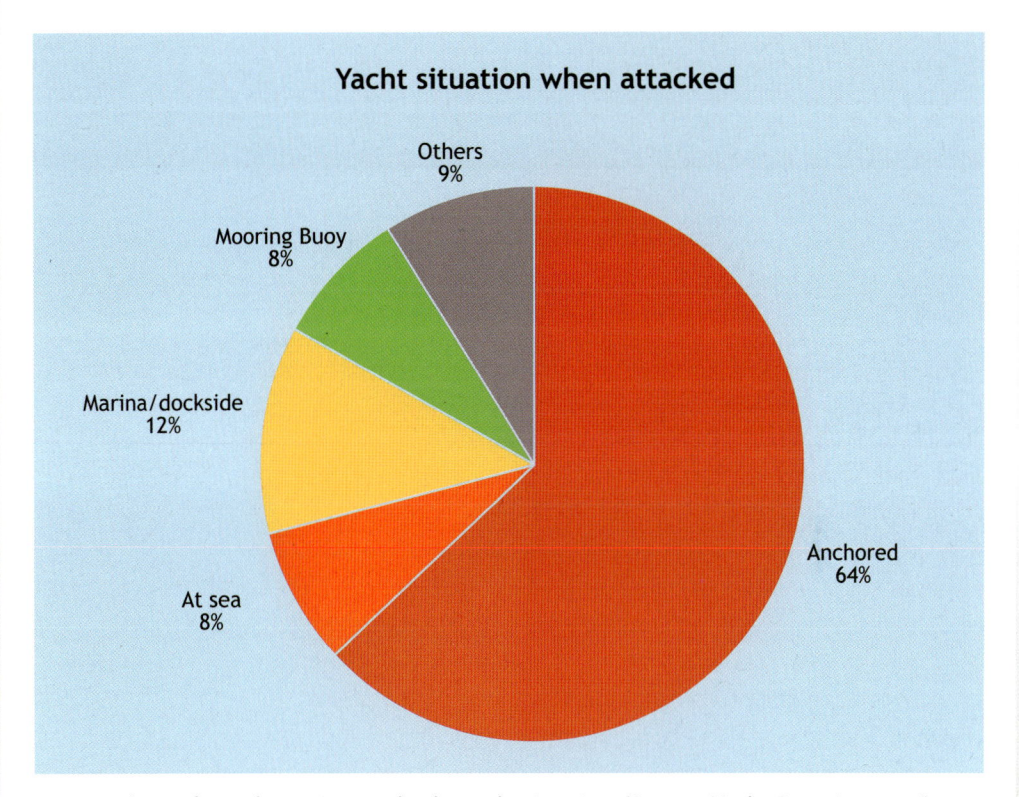

FIG 2.5: *Share of attacks against yachts by yacht situation. (Source: 'Yacht Security 2017')*

2.5. Using the diagram we can see that almost three-quarters of all crimes were committed against ships that were moored or anchored.

Without taking the severity of crimes into account, marinas are surprisingly attractive to criminals. Some definitely have serious security issues (more on that later), but most offer good safety. Compared to the other situations, attacks against yachts at sea are quite rare. Fewer than ten per cent of all reported crimes were true acts of piracy against vessels in international waters. The category 'others' combines assaults against the crew on land, and attacks against grounded boats and boats on the hard in a boat yard.

OPTIMISING YOUR CRUISE: EFFECTIVE RISK REDUCTION

With good data on your cruising region and the general riskiness of yacht situations at hand, the cruise can be planned to ensure carefree sailing and relaxed exploration.

Depending on your situation and interest, there are two ways to ensure this: you either have your route defined and then plan vessel and crew from scratch, or you already have a boat and crew and look for a route that fits their capabilities. While the first situation is usually limited to semi-professional explorations or long-term cruisers at the beginning of their career at sea, the second situation is typically relevant for charter crews, holiday sailors and established liveaboards who want to change their cruising ground without changing vessel or crew.

Equipment and crew capabilities are a bit of a red herring in these equations since equipment can be added to ships easily and, within limits, crews can improve their capabilities or more crew members could be (temporarily) added to improve the general strength of a boat during demanding passages in risky regions.

Before deciding to avoid a beautiful yet more risky area, skippers might thus consider investing in some equipment, preparing themselves and their crew or inviting some more staff on board their vessel.

Unfortunately, there is a large group of skippers who enter rather demanding waters without being prepared. These account for the larger portion of the yearly reports concerning yachts that were successfully assaulted by criminals.

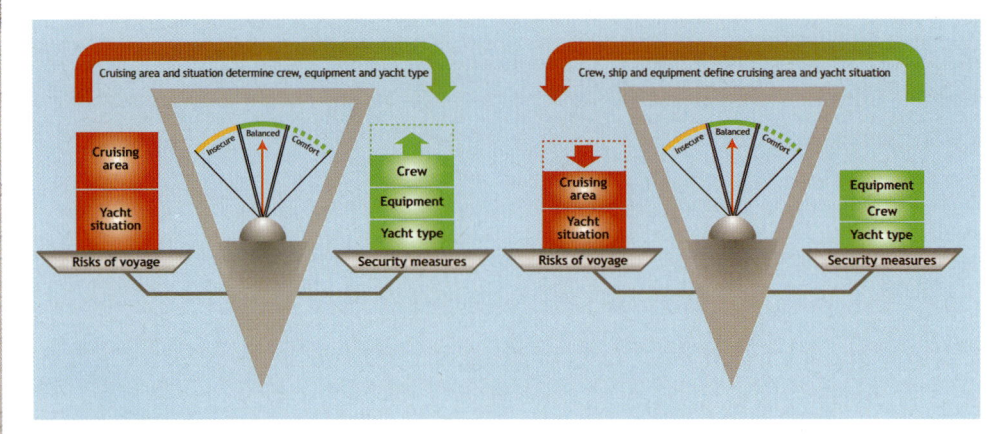

FIG 2.6: Options for a relaxed cruise – adapt risks or security measures so that they balance each other.

PLANNING YOUR CRUISE **47**

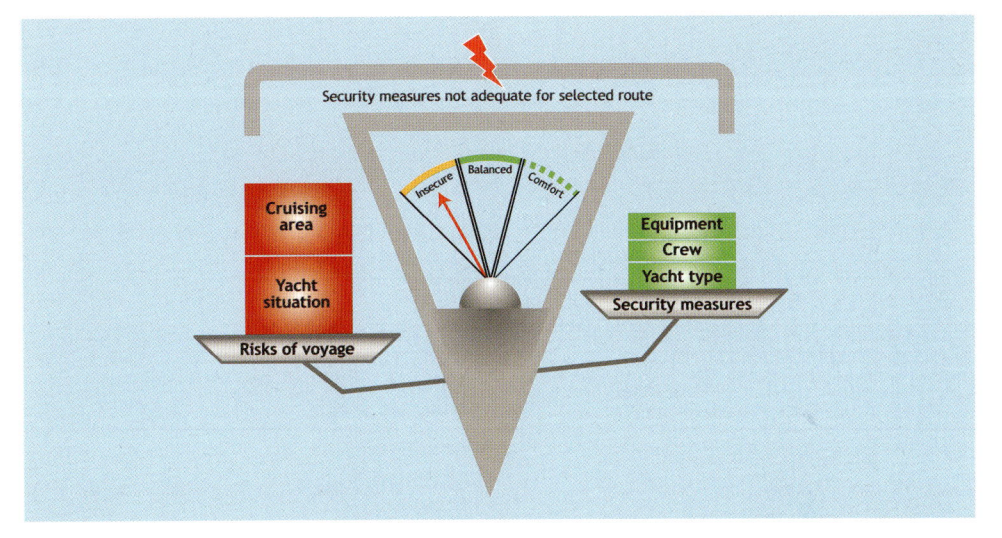

FIG 2.7: *An insecure set-up, usually caused by crews visiting areas with a risk level that is too high for their vessels, crews and equipment.*

In most cases, the skippers overestimate the safety of common cruising destinations, such as some Caribbean Islands. A few reports reveal tremendous naivety when travelling poorly prepared to crime hotspots, such as the bay of Port-au-Prince in Haiti or the waters around Basilan and Jolo in the Philippines.

By contrast, in the last five years there have been extremely few reports about well-prepared crews on adequately equipped boats being successfully attacked or injured by criminals even in Risk Level III regions.

Let us, therefore, take a look at the preparations that skippers could undertake to ensure that their cruise is safe and relaxing. The next chapter will cover the ship type, equipment and crew.

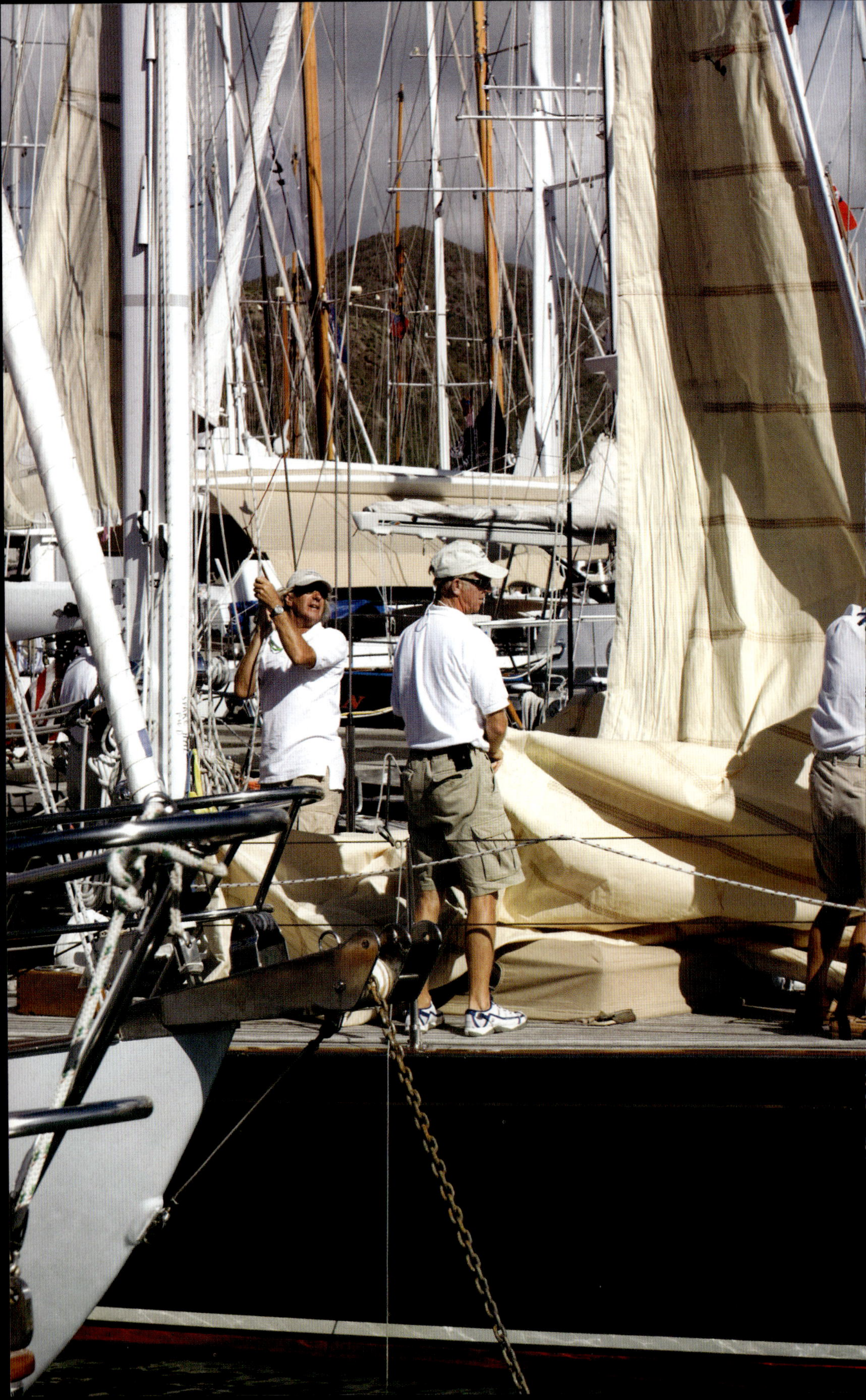

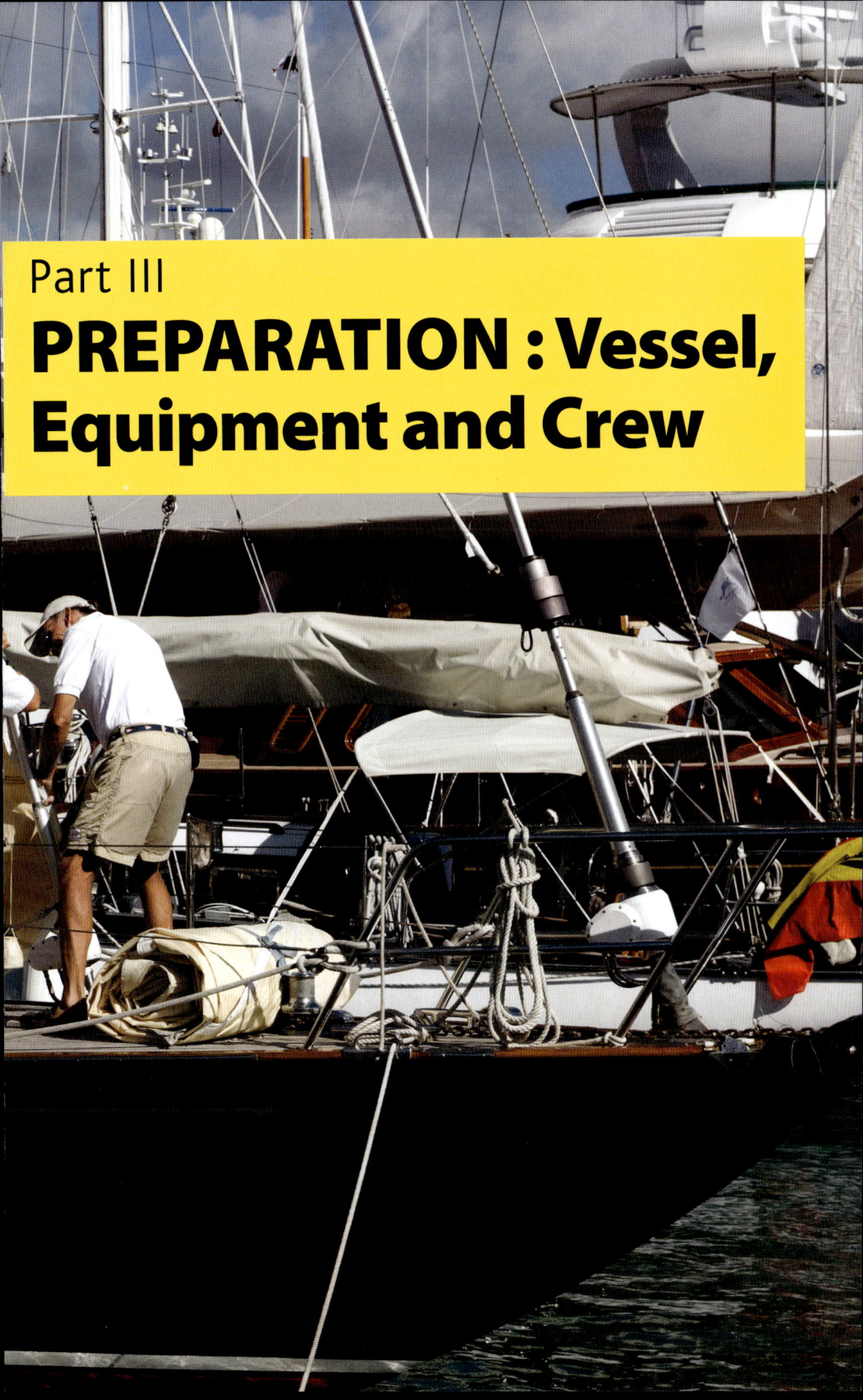

Part III
PREPARATION : Vessel, Equipment and Crew

SHIP LAYOUTS AND SECURITY

> Hull shape, building material and the general layout of a vessel determine its basic ability to withstand or at least impede a criminal attack. However, a typical cruising vessel's qualities only play a role in helping the crew to detect boardings in quiet anchorages by forcing the assailants to either make noise or enter through a few available choke points. During an attack led by determined, gun-wielding criminals, most pleasure yachts do not offer any serious protection.

Boat types and layouts perform very differently in a crime situation. In the extremes, you could compare a warship with a cosy daysailer: one is heavily armoured with high freeboard and covered passages for safe on-board manoeuvring, while the other shows a low deck, open transom and building material that can be punctured with a knife.

As this book focuses on relaxed sailing rather than on how to transform your boat into a floating fortress, I won't give details on how to change its layout and build. However, some vessels do provide a safer environment than others, thus providing a bit of weight on the 'security' side of the balance. Consequently, this section aims to inform readers about how to judge their vessel's strengths and weaknesses.

When assessing your ship in terms of security, you may consider three aspects: the freeboard and the shape of the outer hull; the building material and its ability to resist attack; and the general layout, which may or may not offer cover or even a makeshift 'citadel'.

Hull shape and freeboard

The higher your freeboard, the more difficult it will be for boarders to actually climb up to your boat without being noticed.

A tall and trained assailant starting their climb from a high-sided boat will be able to pull up to your deck even if your sides are higher than 2m (6½ft). However, it will be difficult to do so silently. First, he will need a boat's gunwale to use as a 'launch pad'. If you take a look at Fig. 3.1, you will notice that this sort of boat will most likely bang against your hull if it comes too close. Without someone maintaining some distance between the boats, a single attacker's vessel will also often slide 'under your side', making the climb very difficult.

Next, he would need to initiate and maintain his upward movement without touching your hull by using his feet as support to avoid noise. Even qualified parkour athletes would have a hard time performing such a feat. Finally, he would need to climb over your lifelines and take some steps on your deck before reaching any valuables.

Some vessels have lower freeboards, because they have a smaller volume or have more of their structure below the waterline. As soon as your freeboard is lower than 1.5m (4.9ft),

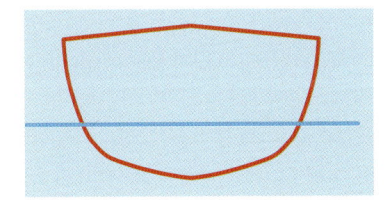

FIG 3.1: Typical cross-section of a cruising yacht.

assailants approaching with boats have an easy time boarding. Add the height of their gunwale and they might end up with just 0.5m (1.6ft) to climb; some attackers could almost step over by holding on to a stanchion while raising one foot high up on your deck.

Many modern boats have at least straight or, more popular today, open transoms with bathing platforms that invite boarders to enter. The same holds true for many multihull layouts: most multihulled boats (especially modern cruising catamarans) feature bathing platforms and comfortable stairs leading up to the deck.

By contrast, some classic designs inspired by old-school racing boats feature overhangs that are very difficult to overcome unnoticed.

However, the benefit of such hull designs will beome less important in windy anchorages or during an attack at sea. Cruisers must be aware that hardly any pleasure vessel's design will offer true defensive capabilities when attackers do not care about stealthiness.

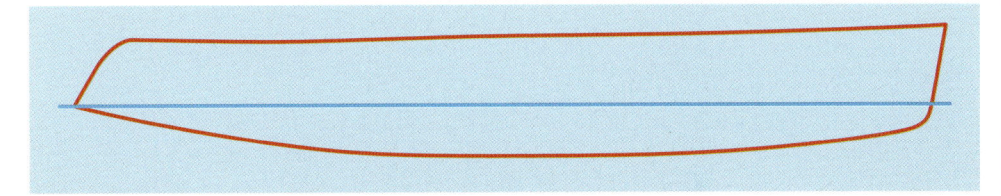

FIG 3.2: Modern hull shape that will most likely feature a small bathing platform at the transom.

When assessing your vessel, just think a little about where a boarder would try to enter. Criminals are usually (but not always) looking for the easiest, most silent access. Try to identify which route they could choose and focus your deterrence efforts, alarming devices and entry barriers there. These should also be the first places you check out if you are woken by strange noises during the night.

In the case of classic boats, offenders in smaller boats would most likely try to board over the sides, since the overhangs at the stern are more difficult to climb. On most modern

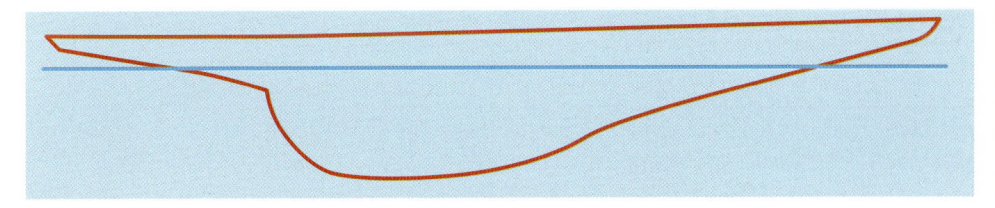

FIG 3.3: Classic racing-hull design featuring long overhangs and a hard-to-climb transom.

boats, you could expect boarders to enter via the bathing platform and transom when attacked at anchor. In the rare event of a boarding attempt while cruising, the sides might be a more likely access point since it is easier for an attacking boat to stay close (or even be attached) at the side while boarders try to climb up.

All of the above is relevant when assailants arrive in small, open boats or as swimmers. In the past, such boats and swimming were by far the most common tactic used by attackers to approach their victims in anchorages and mooring. Larger boats, such as 60ft fishing vessels, have higher sides than most yachts and therefore provide more options for the attackers. Such boats have occasionally been used during offshore assaults, though less often during attacks made at anchorages (because big attacking boats are very conspicuous and noisy). If you are approached by such a vessel, you have to expect to be boarded from any position regardless of your boat's hull design.

FIG 3.4: Low bathing platforms are the most likely access points for attackers.

Building material

A yacht's building material is only truly relevant in the most severe attacks featuring gun-wielding criminals or in situations in which you can consider a ramming manoeuvre. Let's first look at situations that involve armed criminals.

Knowing just a little about how your hull's material might perform may help you to decide whether you want to surrender at first sight of a gun, attempt an escape, hide and call for help or fight it out. However, defenders should bear in mind that many handgun bullets and all bullets from modern rifles will penetrate every part of a glass-reinforced plastic (GRP) and aluminium yacht. This is one of the reasons why this book recommends bypassing any region with a recent history of unresolved crimes involving firearms.

To be a little more precise, whenever you see a long-barrelled rifle, be prepared for the fact that your boat will not provide protection against the bullets. If you only see handguns, the chances are a little better. Weaker handguns, such as in .38 revolvers or 9mm pistols, are quite often used in criminal attacks outside of pirate-plagued waters. Your hull may stop standard projectiles from these weapons if the gun is fired from a distance further away than 50m (165ft). When fired from a closer position or even on deck, projectiles from these guns penetrate all parts of a typical cruising boat's hull and superstructure. The sharper the angle at which they hit your boat, the better for you: the probability of deflection increases with sharper angles. The lower down you are in the boat the better, as the hull material usually becomes thicker and the water helps to stop the bullet. Many boatbuilders also increase material strength in the section close to the bow. Consequently, if you would like to protect someone or something from external gunfire, spaces in the front area below the waterline would be best so long as you are not moving towards the attackers.

A bullet from any other weapon (higher-powered handguns and rifles) or rarely used rounds of especially high power in the calibres mentioned above will most likely penetrate any part of your boat from further away and may injure anyone inside.

If you want to prepare for the rare eventuality that you will meet gunmen, you should walk around your yacht bearing in mind the difference between concealment and cover. All objects and structures that hide you from the attacker's view provide concealment. Everything that is actually capable of deflecting or stopping a bullet is considered cover.

As stated above, only very few structures provide cover on a GRP or aluminium yacht. The engine block and battery banks may be the only safe bet. A full fridge, your water heater and the oven are less reliable but better than anything else on board. If you feel daring and own an aluminium boat, you might add doorframes, stringers and the ribs of your yacht's hull to the list. However, these are very slim structures and (in case of hull structures) difficult to identify. To stop a bullet from a typical 9mm handgun, the aluminium sheet would have to be at least 1cm ($^2/_5$in) thick (more for a harder, heavier and/or faster bullet).

Although a steel boat's hull may also be able to stop projectiles from some weaker handguns, most rounds from a rifle will penetrate one layer of metal and injure the crew behind it. Consequently, a steel boat does not provide true cover against rifle rounds unless you are able to position at least two walls between the shooter and yourself. Nevertheless, as most non-professional assailants used handguns in past attacks, many parts of a steel boat provide some kind of cover against such a threat. This gives the crew the option to hide in relative safety while they get the boat moving in order to escape, defend against the borders or signal for help from positions that are fairly protected. However, be very aware that there cannot be certainty about the safety of the position: the ability of a steel plate to resist a handgun bullet is dependent on the bullet type (full metal jacket, partly jacketed, lead only etc), the handgun power and calibre, the angle of the impacting bullet, the shooters' distance and the thickness and quality of the metal used in your boat.

In the very rare situation of a pirate attack while sailing, one desperate defensive manoeuvre could be to ram the assailants' boat (see 'Sailing in pirate-plagued waters', page 130). While it is suicidal to attempt such a manoeuvre against a similar-sized boat made from the same or stronger material, it may be considered as a last resort when attacking smaller boats (such as open skiffs or canoes) in self-defence.

Most modern boats – especially steel and aluminium ones – stand a good chance of surviving a ramming against a wooden or GRP skiff that is shorter than 8m (26ft) without sustaining catastrophic damage to the yacht. Bearing in mind the build of the assailants' boat, it is best to hit the attacker in the forward or mid-section of his vessel to minimise the chance of a collision with his engine. If his tub features an onboard engine, try to go for the front section, as there may be a very robust diesel tank installed in the mid-section. While approaching head on, you are also protected by a lot of material from your boat, as long as you keep your head down.

General layout

It is well worth taking the time at least once to explore your own boat with a defensive situation in mind. While doing so, try to identify the passages that could provide concealment while the crew moves from their bunks towards the suspected boarders (or the place where you would expect a boarder to step on deck first). It is also wise to work out whether you would be able to use your most powerful radio, your foghorn and other emergency signals while remaining concealed or even covered.

Usually, deck saloon yachts offer more concealment while manoeuvring outside because the crew can duck or crawl behind the saloon's walls. Most modern cruising catamarans also feature such a high structure. By contrast, on many modern-layout monohulls without a deck saloon concealed movement is only possible below decks or in parts of the cockpit.

Modern boats feature a layout with multiple emergency exit routes to prevent those on board being trapped if there's a fire. Such designs enable the crew to exit from below decks via at least two independent routes. Usually, the alternative exit leads through a hatch in the front of the boat. It is handy to know about these exit points since you can use them to either flee altogether or flank/surprise any boarders while they try to break into your companionway (or sliding doors in a multihull).

Some larger deck saloon yachts and recycled work boats even feature very robust entry doors that can be effectively blocked from the inside, thus creating a makeshift citadel in which crews can shelter for a while. Even less-robust 'citadels' – locked rooms with some protection against the attackers – have proven handy in some reported assaults. For

FIG 3.5: *Dutch quality on view in this door leading to the robust deckhouse of a Koopman monohull.*

example, one skipper was able to fire a flare gun through an open hatch while attackers tried to break the blocked door to his hideaway. They left after the second flare summoned people from the beach.

Despite rarely being needed, rooms and compartments below the waterline can provide relative safety during firefights. Children in particular or others who are unable to actively participate in a conflict or an escape can be hidden there. These are also good places for the whole crew to be during any rescue attempts by security forces after the yacht and crew have been kidnapped.

SECURITY EQUIPMENT ON BOARD

Adding security equipment to a yacht is a versatile method of supporting crew in more risky waters, although most is only necessary when cruising Risk Level III regions. While anchoring in such areas, the majority of skippers can greatly increase their peace of mind if they simply haul and lock their dinghy, lock their boat while asleep and invest in some cans of pepper spray and a compact battery-powered infrared alarm. Should there be an attack, crew members can also improvise using standard

> equipment. Items that support ventilation will greatly increase comfort in warmer countries.
>
> It is important to remember, however, that while more sophisticated equipment increases security, it also puts a strain on the budget. That said, skippers who intend to spend prolonged periods in Risk Level III regions or even plan to sail through Risk Level IV waters should at least consider some of the more elaborate equipment presented in this chapter.

One of the easiest ways to get a good night's sleep is to add effective security equipment. Some of this is incredibly cheap yet outstandingly effective against many threats. This book covers equipment that might be helpful for pleasure yacht crews and will place a special focus on items that have been successfully used in the past. However, we will also propose some that don't appear in crime scene reports.

Depending on the risk area and the type of crime that may be encountered, different equipment will be useful. For instance, you will not need trailing nets to foul a pursuer's propeller in the Med, nor will you need a burglar alarm while passing through the Gulf of Aden, unless you plan to stop.

To help you to select the right equipment for your boat and trip, the 'security quick score' (see the key in Fig. 3.6) shown alongside each item shows the relevant risk level, yacht situation and a security score. The security score is defined by judging an item's usefulness in actual security situations (or their avoidance).

In the following sections, the items are categorised according to their role before or during criminal attacks, as detection, alarming deterrent, defence/resistance and denial. Some types of equipment have dual roles (think of a weapon that both deters attackers and supports resistance). If possible, the primary role is used for allocating an item to a category.

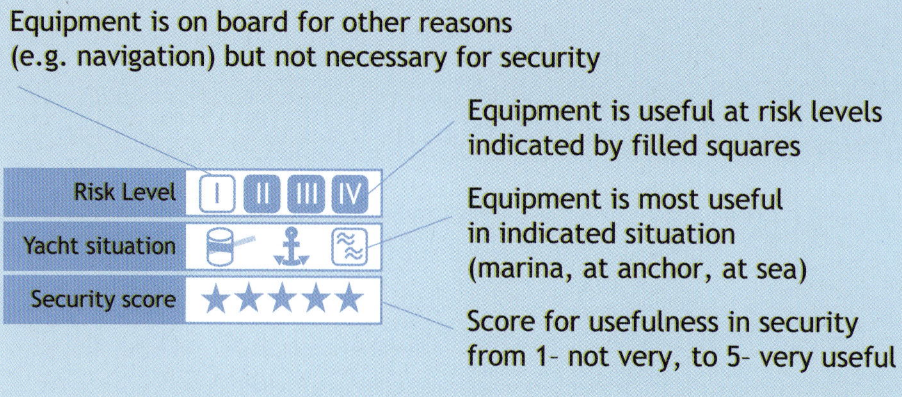

Equipment is on board for other reasons (e.g. navigation) but not necessary for security

Equipment is useful at risk levels indicated by filled squares

Risk Level	I	II	III	IV
Yacht situation				
Security score	★	★	★	★ ★

Equipment is most useful in indicated situation (marina, at anchor, at sea)

Score for usefulness in security from 1- not very, to 5- very useful

FIG 3.6: The 'security quick score' helps to identify the right equipment for your cruising plans.

Denial: locks and theft protection

Padlocks and chains

Risk Level	I	II	III	**IV**
Yacht situation	🛢	⚓		
Security score	★★★★★			

Padlocks are simple and, if they are the right models, effective items in a yacht's security arsenal. They are used to secure dinghies, deck equipment, doors and cockpit lockers. Whenever buying a padlock, skippers should only consider ones that use the highest-quality material. The locks will be frequently exposed to seawater and anything but the best metals will rust quickly. Consequently, even better-quality padlocks intended for onshore use will be impossible to open after a short while. For high-value items (such as a dinghy) or in very sensitive areas (such as the companionway), you might consider the highest class of padlocks, offering a hermetically sealed interior and 100 per cent water and dust resistance.

Skippers should remember to service the lock itself with a protective lubricant such as WD-40 at least once a month to ensure that they will be able to unlock it at any time.

It is surprising how fast a determined attacker can open even a high-quality padlock with an accessible shackle in just under one minute. For the best locks with a $^3/_8$in (1cm) shackle, they will need less than two minutes. However, this can't be achieved with the easily concealed tool preferred by thieves; to force such a lock, the offender will need a 42in (1.6m) bolt cutter and sufficient room to support one handle of the tool on a hard floor. A lot of jumping on the other handle is then required to build up the force that will eventually break the lock. Most thieves tend to both avoid carrying a conspicuous tool and performing an attention-grabbing 'handle-dancing show' for potential witnesses at the local dock.

As a result, a yacht's equipment is very safe in the short term when it is secured with a $^3/_8$in lock of highest quality, especially when this has been positioned in a place that does not offer a solid surface for leveraging force. If the lock is used to secure a companionway on a boat that is not manned for a longer period of time, a ½in (1.3cm) shackle is almost indestructible when using a bolt cutter.

Having decided what size padlock you need, you should now consider the material and construction. Higher-quality through-hardened steel, such as boron carbide, offers better protection against bolt cutters compared with regular stainless steel, and the most secure padlocks feature raised shoulders that minimise the area where bolt cutters' jaws can

FIG 3.7: High-quality lock with raised shoulders. (Photo: Assa Abloy)

be applied, thus further protecting the shackle.

One final consideration regarding locks for high-value items is their vulnerability to picking. With just a little practice, a simple lock can be picked with a pair of hairpins in under 30 seconds, using methods that are readily available on the internet. To prevent picking, higher-quality locks employ movable parts in their keys or alternative designs to the very widely used and easily picked pin tumbler locks. Some of the alternatives would be disk tumbler, tubular or cruciform locks.

Chains are equally important. While every stainless-steel chain from your favourite chandler with a diameter larger than ¼in will be difficult to cut with simple tools, security-aware skippers might want to look for some special features to protect their valuables. Generally, chains with square cross sections are considered more difficult to cut than rounded models. This is because the flat surface distributes loads from a bolt cutters' jaws more evenly. Better chains carry the label 'through hardened' as opposed to just 'case hardened'. The latter means that only the surface has been treated to be more resilient, while the former implies a material that has been hardened from surface to core.

Finally, when buying a chain and lock, it is a good idea to measure the links' interior diameter and to buy a lock that will fit through.

Why not use wires? Wires are much easier to cut with simple, easily concealed tools such as wire cutters. Due to their build, they are also more prone to corrosion: seawater is difficult to flush out of the groves between the single strands, leading to rust in even higher-quality stainless-steel types. Ones with a plastic tube are even more prone to corrosion as seawater is trapped in the spaces between the tube and the metal.

FIG 3.8: *Security chain with square links. (Photo: PEWAG)*

Dinghy and outboard protection

In the 'Yacht Security Study 2017', dinghies and outboard motors accounted for 40 per cent of items publicly reported stolen by thieves in the last five years. An example: in Germany, between 900 and 1,300 outboard engines are filched every year (this figure includes engines stolen from sheds, boats, yards and trailers, and both coastal and inland waterways). This is an unexpected figure given that Germany is a typical 'Risk Level I' country. At the same time, outboards and dinghies are some of the most important pieces of equipment on a boat that is outside regions where there is a good marina infrastructure.

Surprisingly, in many cases analysed in the study the dinghy was described as being fixed to a solid object and the outboard secured, usually with a cable and padlock. However, in most cases skippers were also lazy: the tender was still in the water in all of these cases. It is, therefore, probably safe to assume that most lightweight locks and cables will be of limited use against today's well-equipped thieves. In some situations, the only truly safe places for a dinghy at night are at a surveyed dock or off the water, preferably locked to your boat.

Consequently, the best defence against theft in the anchorage is not necessarily a lock, but rather a system that lets you easily hoist your dinghy in the evening.

Pulling up your tender: dinghy weight, davits and improvised hoists

Risk Level	I II III **IV**
Yacht situation	🛢️ ⚓
Security score	★★★★★

When thinking about how to hoist your dinghy every night, skippers shouldn't just focus on davits and pulleys; instead, it would be wise to start by considering the weight of your tender. This includes the dinghy, engine and engine fluids as well as oars, anchors and tools you store in it. The easier it is to hoist up the tender, the more you will actually do it.

We have learned this the hard way, and have cursed many nights while pulling up our 130kg (286lb) dinghy/outboard combo to our catamaran's davits. By contrast, one of our neighbours once happily lifted his using just one arm and a simple pulley. Unlike us, he had considered performance and weight aspects and invested in a 3hp electric engine that

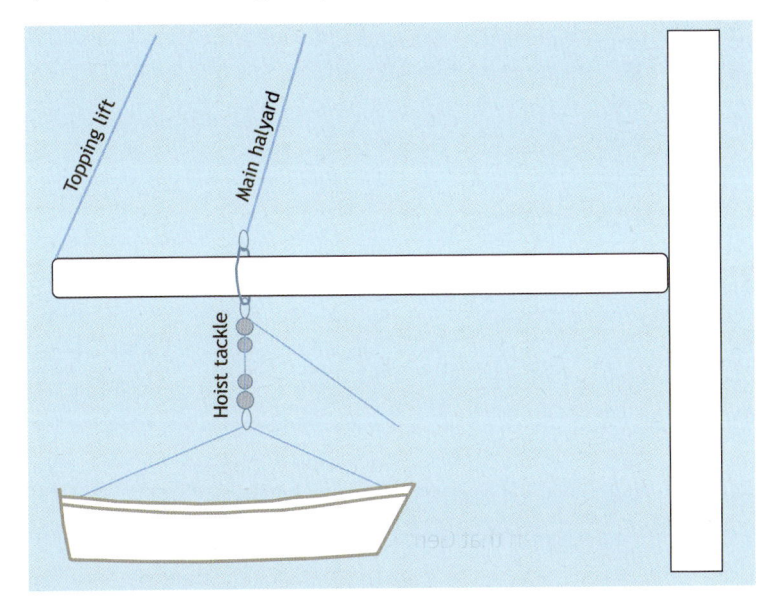

FIG 3.9: *Improvised dinghy hoist to secure the tender away from the water.*

was attached to a high-pressure inflatable floor tender. He took the outboard's detachable battery on board while stepping over to his boat, which meant that the remaining weight to lift was a mere 41kg (90lb). A lighter tender will also require lighter davits and hoisting tackle, reducing the overall weight of the yacht.

When working out how to hoist your dink every night, you should start to think about systems that can help you in the process, unless you plan (and are able) to lift the tender. Most catamarans come with factory-equipped davits, which solves the problem right away, and some monohull models specifically targeting the long-range cruiser market (such as Allures or Ovni Alubat) have a gantry arching over the bathing platform, ready for davit installation. However, most production monohulls do not have this feature, in which case skippers must either custom fit gentries or davits or use improvised methods to hoist.

A common practice is to use the boom and main halyard to lift the tender to the level of some fenders at the side or to the stanchions. In this set-up, the boom does not carry any weight, as this might lead to bending and damage; instead, the boom helps to establish a fixing point outside of the lifelines and over the water. A loop of rope around the boom becomes central, and the main halyard is attached on its upper side to take all of the weight. The tackle to hoist the tender is shackled to the lower side of this loop.

With the boom rigged over the side, the dinghy can now be positioned below the fixing point – attached and hoisted without much hassle. If there is swell or wakes, the dinghy should also be fixed to the side by extra lines.

Depending on the type of rig (mast furling, boom furling or lazy jacks), the way to ensure that the halyard supports the boom (to prevent the spar from bending), and the way in which the loop is slung (to prevent the sail being damaged), will vary. Skippers should also check how much weight their halyard and masthead tackle can actually support.

Some skippers choose to unmount their outboard engine and simply pull the dingy up, using their main or spinnaker halyard to lower it on deck from there. This may be a good tactic if the dinghy is light and electric winches are installed.

Dinghy locks and chains

Risk Level	I	II	III	IV
Yacht situation	🛢	⚓		
Security score	★★★★★			

When analysing reports of dinghy theft there was a surprisingly high number of reports of tenders that were secured with padlocks and cables to docks or yachts (not hoisted up). Today, offenders seem to be well equipped with bolt cutters in regions notoriously known for dinghy theft. Even heavy braided $3/8$in steel cables can be cut in seconds with 14in (35cm) portable cutters, which are easily concealed in Caribbean-style surf shorts.

For this reason, it is a good idea not to rely on cables at all. Instead, skippers should consider buying stainless-steel chains. Dinghies remained untouched when secured with chains with diameters of at least ¼in. Theoretically, 8mm of stainless steel will resist cutting profoundly longer, although their weight becomes a problem. Hardened chains with a square cross section in particular are more secure than the traditional mariner's round-sectioned anchor tackle.

Measured from the dinghy's gunwale, the chain should be at least 2.5m (8ft) long. This ensures that there is sufficient chain to wrap around fixed objects and that the tender can be moved away from the dock to allow others to attach their vessels.

Some length could also be considered inside the tender. This can be used to fix both the motor and the fuel tank with one piece of chain. If one end of the chain is fixed inside the dinghy, it will rub against the PVC or Hypalon of its air chambers, possibly damaging them. Experienced cruisers sometimes protect their dinks by simply gluing an extra layer of the building material on the hull where the chain usually runs over any vulnerable parts.

When attaching your chain to a fixed object on the dock, try to use places that are below the platform level (supporting posts, etc) to ensure that a thief cannot use the platform or parts of the dock's construction as a support for one handle of his bolt cutter. Cutting stainless-steel chains without such a support is only possible when using very conspicuous, long and heavy tools.

Even higher-quality stainless-steel chains will start to corrode and produce ugly stains in the tender when left in puddles of seawater within the dinghy. To reduce this, as well as the nuisance (sound and damage) caused by a length of loose chain on the dinghy floor, some skippers use draining plastic boxes or baskets as chain storage in the tender. Doing this means that the chain will not slide about in choppy conditions and it is easily rinsed with fresh water.

Outboard engine locks

Risk Level	I II III IV
Yacht situation	
Security score	★★★★★

An unsecured outboard is very attractive booty for thieves in many cruising areas. As far as they are concerned, there is no big difference between it being fixed to the dinghy's transom or a yacht's mounting bracket. To prevent a quick hit-and-run tactic, several methods have proved effective. Many motor models feature a hole in the mounting bracket that is intended to house a safety line when the engine is moved from a tender's transom. Usually, it is drilled through a strong metal section of the bracket and could be used if you wish to secure the motor to the tender with a strong padlock and chain. The chain could be a separate piece or part of a longer section that is also used to lock the dinghy to the boat or a dock when left alone.

Using just a bolt through the holes in the motor's mounting clamps is cheap and simple, but is not an effective solution: the clamps' handles are usually made of plastic that can quickly be broken with a hammer or bolt cutter without risking damage to the engine –

and indeed this is a common tactic.

A better solution is so-called slot locks, which comprise a tube that is slid over the heads of the motor's mounting clamps, thus protecting them. The tube is then secured with a lock to allow access for the owner. The best versions feature high-quality locking mechanisms and are more resistant to rust.

Heavier outboards that are rarely taken off the tender can be secured with bolt locks that protect at least one of the bolts securing the motor to the boat's transom, making the whole construction semi-permanent. The set-up is very secure but not very convenient on boats that need to detach the outboard to reduce weight on their davits.

FIG 3.10: *Effective DIY slot lock. (Photo: N Langford)*

Anchor chain (extra long)

Risk Level	I	II	III	IV
Yacht situation	⚓			
Security score	★★★★☆			

Readers might be surprised to find anchor chains in a book on yacht security. The reason for including them in a list of relevant equipment is their impact on two factors:

Normally, a longer chain allows a crew to anchor further away from the beach. In most conditions, each yard away from a seedy beach will make a vessel safer from criminal attacks originating from the shore. When forced to stop at an anchorage that simply does not feel good, skippers are well advised to remain as far away as possible. Distance will discourage swimmers and paddlers alike and it will give crews longer to react whenever engines approach at night.

Longer chains also make more anchorages available – quite a few places in the world are simply not accessible if you only have a standard 45m (150ft) charter-boat chain. Even when carrying a 64m (210ft) one on our catamaran, we were envious in many cases when we saw other boats that had invested in chains and cables longer than 90m (300ft) drop their hook in beautiful anchorages with great holding at 27m (90ft). While the safe chain-to-depth ratio of 3:1 limited us to depths not deeper than 20m (66ft), they could drop their hook in anchorages with great holding at 30m (100ft). In less-favourable conditions with strong winds or bad holding we had to anchor even closer to the coastline – not great from a security perspective – whereas crews with the longer chains had the option of using back-up bays that were unavailable to our yacht. Depending on the topology of the area visited, more than half of the anchorages are off limits if the yacht is carrying an anchor chain measuring less than 90m (300ft).

Companionway security

Risk Level	I	II	III	IV
Yacht situation	🛢		⚓	
Security score	★★★★☆			

When looking at reports of sleeping crews who have been surprised by burglars and robbers, it becomes evident that the greatest security risk is the crew itself: in many such cases, the companionway had been left wide open or at least not locked. Intruders have therefore been able to easily sneak down the companionway to wake those below at knifepoint. As in many situations, security concerns had conflicted with comfort. Since many boats come without air conditioning and devices to secure an open hatch, entry points had been simply left wide open to reduce the temperature in the cabin. Had the victims slept behind locked doors – as most of us do at home – many thefts and most confrontations would have been avoided.

FIG 3.11: *Osprey Marine's hatch latch. (Photo: Osprey)*

Unfortunately, the companionways of many ships built in large boatyards are secured insufficiently. Although insurance companies accept them as proper protection, a criminal can generally open them in a short time with a lever, a bolt cutter or by picking the simple locks. Obviously, it is important that the insurance company will cover the damage, but having some of the damage compensated is not truly relevant at the moment when a sleeping crew is below decks while robbers try to get inside, or when the yacht is cruising in areas without access to spares for items stolen from the cabin.

FIG 3.12: *Simple yet effective stainless-steel bars in the companionway. (Photo: www.atomvoyages.com)*

Major weaknesses are the simple pin tumbler locks, the thin

and sometimes multi-panelled washboards (which are easily kicked in), and the weak connection between the top sliding hatch and washboards established by the locking mechanism. Due to the thin materials used, there is a lot of scope to simply 'wiggle' the lock open in some set-ups.

Before investing in vamped-up companionway security, however, a quick consideration of its main goals is advised. What should be protected in which situation? Will the weather be hot? Most cruisers want to ensure that their boat's interior is safe while they are on board, as well as during the short periods when they are exploring the shore area. The security should also be able to withstand attacks in populated anchorages or marinas if a boat is left for a couple of days while the crew explore the visited country's interior or take a short trip home. Consequently, skippers need a solution that buys them some time to react and call for help should they be attacked while below decks, and that deters offenders before they even try to break in. It should also be capable of withstanding concerted and prolonged break-in attempts since a delay could potentially cause an attacker to give up.

The aftermarket offers some locks that are superior to many systems installed on production-model yachts. The best ones offer simple mechanisms that can be secured with a robust padlock from the outside and with a quick-release blocking device from the inside. All parts should be bolted through the panels that they are installed on to prevent removal with a crowbar.

Installing a strong, state-of-the-art lock on flimsy doors, however, could motivate any attacker to simply kick in your washboards. These are often made from thin and easily broken material. Some plastic versions are so soft that they can simply be pushed out of their rails to remove them. While this is sufficient in safe regions, owners of such set-ups might consider having a replacement made from more a resilient material, such as plywood or aluminium, which they can carry on board and install when entering areas with a higher risk level. These replacements are simply cut from the desired tough material using the original washboards as a template.

During the summer and in areas with generally higher temperatures, ventilation becomes key for enjoyable sailing. In these conditions, a stainless-steel frame with some bars can be built at a metal workshop. This will subsequently replace one or all of the plywood or aluminium washboards.

Sliding door security

Risk Level	I II III IV
Yacht situation	🛢 ⚓
Security score	★★★★☆

On catamarans, sliding doors are the typical means of closing the passage from the cockpit to the saloon. In most designs, the large door interlocks with the frame of a sliding window with a lug turning with the

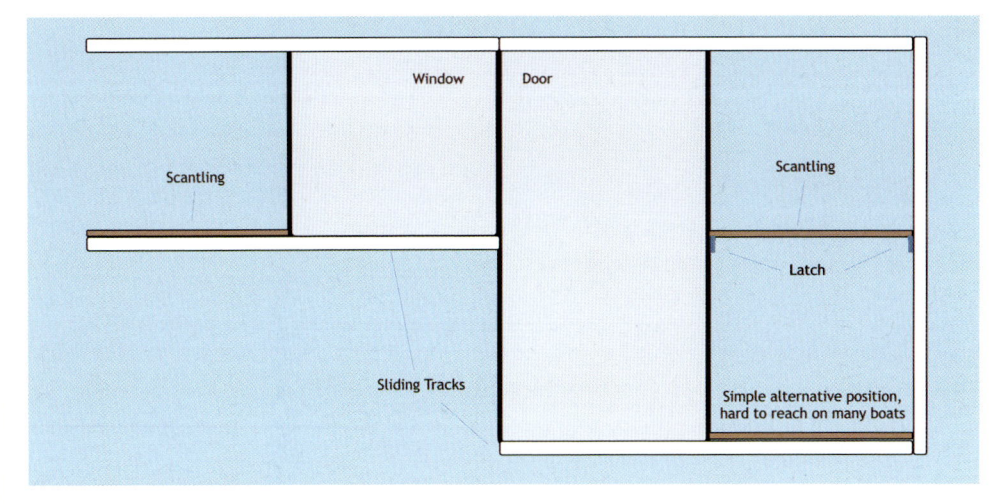

FIG 3.13: *Scantlings positioned to prevent the opening of sliding doors.*

key. Many of these set-ups – especially on older models – are very easily opened, even if properly locked.

We were lucky and shocked when we had to break into our own Lagoon 400 (built in 2011), having lost our key on a shore excursion. All that was required was a determined upward shove on the smaller window (there is a lot of play in the vertical plane) and the locking mechanism could be 'wiggled' open. The window was then slid aside and we unlocked the door from the inside by simply reaching through the opening. Even without turning the fixed indoor key, we could have pushed the door panel open using mild force without damaging it. The whole procedure was completed in less than 30 seconds by a skipper who was absolutely not familiar with break-and-entry tactics.

To effectively prevent the bad guys from opening sliding doors while a crew is sleeping inside, a very simple solution involving two scantlings or bars made from wood or aluminium can be used. These should be cut to the right length and positioned so that they block both the window and the door in their sliding tracks.

If these are cut a little shorter, the door or window could be left open a crack for ventilation. If this solution is used more often, it would be a good idea to install simple supporting latches on both the door and the frame to make the scantling easily accessible for the crew should the cabin need to be exited in a rush.

This set-up will not work if no one is inside. The only way we can currently think of to secure sliding door/window combinations with a lot of play in the system is to install a hardened steel U-bolt through both the window and the door and connect the two with a good padlock. The bolts need to be securely countered on the inside of the door and the window. Since the lock cannot be operated from the inside, this is only appropriate in instances when no crew members have been left indoors, since they would be trapped in an emergency.

Hatch security

Risk Level	I II **III** IV
Yacht situation	🛢️ ⚓
Security score	★★★★☆

In terms of the main hatch, the major risk occurs when the crew leaves the top hatch open or not properly latched in the hope of catching a cool breeze while sleeping below deck. Several reports decribe incidents where borders used open hatches to entre anchored yachts and quickly overwhelmed the sleeping crew by using the element of surprise.

However, when properly locked from the inside, modern hatches are fairly safe. There are ways that robbers or burglars can get them open, remove the seals or use heat to melt the plastic, but these are loud, lengthy procedures that deprive criminals of the chance to surprise a sleeping crew.

Nevertheless, crews like to sleep with some air circulating in the boat. To allow for this, several solutions can be considered.

One option would be to install models that feature a built-in hatch vent or to retrofit existing hatches with such a system. Due to the fact that these vents must also protect against water, they unfortunately do not allow a lot of air to get into the boat.

FIG 3.14: Hatch vent before installation. (Photo: Lewmar)

Some hatch models offer a venting setting for the hatch lid. With such systems, you can open the lid just a little and lock the handles safely in place. This will prevent any attacker from surprising the crew by jumping at their beds from above. Nevertheless, the opening is still quite small and the sleepers will only feel a breeze if the hatches open to the direction from which the wind is coming and a decent breeze is blowing to push some air through the narrow slit between the frame and the lid.

Some skippers place fixed stainless-steel bars or grates with heavy screws under the actual hatch lid. These are effective in preventing burglars from entering. However, in addition to creating a general prison-like look and feel, such constructions also block an important escape route in cases of emergency, such as capsizing or fire. Accordingly, we strongly discourage the use of such fixed devices.

A more suitable variation uses padlocks that can't be opened from the outside but can be opened from the inside.

FIG 3.15: Battery-powered fans help to keep the hatches safe and cool. (Photo: Kaz/ Honeywell)

The operators of such systems should ensure that they have the key for the locks close by, and they must subsequently assume that they will have the time and clear mind required to find the key, open the lock and remove the grate if there is an emergency. Looking at the performance of humans under severe stress and adding smoke, heat or gushing water to the scenario raises some serious doubts. However, one option to mitigate this would be to remove the grates before raising the anchor for a longer passage, thus reducing the risks.

Low-energy fans are also a possible venting solution. If their consumption is taken into account when assessing the energy balance of a boat, they could run during the night (or at least during the hot late-evening hours). If smallish battery banks prevent skippers from using such ventilators, they could still use products that run on standard household batteries for those extra-hot nights. Such fans are cheap to buy, are said to last a few nights on a single set of batteries, and keep the crew cool while the locked hatch keeps burglars outside.

Detection

Detection plays a major role in yacht security while sailing through high-risk waters. Consequently, the single most important action – especially in pirate-plagued waters – is continuous vigilance using a good pair of binoculars. Substantial investment in night vision, thermal imaging or image-stabilised magnification can of course achieve further improvements, but unless the plan is to remain in high-risk areas for a prolonged period time, the extra security may prove less useful than hoped. In many cases, a highly alert lookout and the presence on board of tools for proper navigation provide a very good basis for security.

Binoculars

Risk Level	I	II	**III**	**IV**
Yacht situation	⚓	≈
Security score	★★★★★

Since a set of binoculars is needed for navigation on any seagoing boat, most crews will automatically have some means of spotting suspicious vessels on the water. However, when passing through Risk Level III or IV areas, a very high level of vigilance is required in order to ensure safe passage. As discussed in 'Detecting the threat: the effective lookout' (see page 150), good binoculars are essential for safety in such regions.

It is generally best to choose glasses from the upper-mid price range and upwards since this should help ensure you get great quality for your money but don't overspend; at a certain point, the upsides of the most expensive models are described by many testers as being subjective. Skippers planning passages in higher-risk level areas should bring a good set of 7x50 binoculars with strong low-light performance. All other features are helpful but not essential.

FIG 3.16: *A keen lookout with a pair of binoculars is the most effective security measure on board.*

Marine binoculars often come with a x7 magnification, which is a good compromise since they offer large images, a wide field of view and the ability to stabilise the image by hand on a moving platform (such as a yacht). Consequently, x7 is a good magnification for spotting targets and identifying them at a reasonable distance.

Higher magnifications are better for identifying targets once spotted, although it is difficult to keep the object motionless, and therefore crisp, without technical assistance such as electronic or optical image stabilisers. These add to the price substantially. If the budget allows for this, then these higher-spec binoculars with x14 magnification could be a very useful addition to your watch as a back-up tool to identify a threat and its armament after it has been spotted with a good x7 model.

Whichever you go for, it is important that you pay attention to the lens size. The larger the lens, the more light can be collected and transmitted to the spotter's eye. Consequently, in order to detect a suspicious approach even in lower-light conditions, the lens diameter of a good spotting glass should not be smaller than 50mm.

A built-in compass is very helpful when communicating the exact location of a spotted target to other crew or vessels and when comparing a target with other onboard detection systems, such as an Automatic Identification System (AIS) and radar. A proper damping is a valuable trait when trying to get your bearings in rough seas. Some budget models feature digital compasses that display the bearing with black digits on your field of vision. However, these are difficult to read against the dark surface of the sea and totally useless in the twilight hours. Effective illumination of the compass – preferably red light to avoid interference with the watchmen's night vision – is essential in low-light conditions.

Some glasses also offer a reticule that enables you to estimate the distance to a spotted object. This is a helpful feature when trying to measure the range to known and stationary targets (such as lighthouses) in a low-stress situation. However, using these reticules to estimate distances of objects that are unknown in size and potentially closing in rapidly is very difficult for an untrained crew under stress. Rough estimates relevant for a yacht's security can be undertaken without such devices.

Bearing compass

Risk Level	I	II	III	IV
Yacht situation				≈
Security score	★★☆☆☆			

Bearing compasses are considered standard equipment on any seagoing yacht. If the binoculars on board are not equipped with one, a bearing compass should be readily available for the crew on watch. In yacht security, they are mostly important while actually at sea. Illuminated models are superior to those that merely offer a phosphorescent dial since the latter usually needs to be 'recharged' by a flashlight before it can be used in the night, which gives away your position and puts the crew on guard at risk of temporarily losing their night vision.

Night vision goggles

Risk Level			III	IV
Yacht situation			⚓	≈
Security score	★★☆☆☆			

Night vision goggles (NVG) are devices that produce visible images in levels of light approaching total darkness. In very simple terms, the technology amplifies both visible and invisible (infrared) light and displays it on a screen. In many cases, the image is depicted in monochromatic green shades. Most systems offered look like binoculars or monoculars and are used just like them. Some miniaturised models can be strapped on the user's head, making them hands-free.

Most NVGs are very sensitive to bright light sources. Consequently, their usefulness diminishes quickly in mixed-light environments such as anchorages close to a developed shore.

They are mostly used while travelling at sea. However, NVGs might also prove handy if a crew decides to be on watch at anchor or if they want to recon suspicious noises in a dark anchorage.

The systems offered on the market vary markedly in terms of image quality. One indicator offered by the producers is the generation of the system, which describes the general technology employed in the goggles. Another figure – the figure of merit (FOM) – indicates the image quality, regardless of the technology used to produce it. The FOM grades the image according to the system's noise. The higher the FOM, the better the quality. Today, a FOM of above 2,500 is the best available, albeit very expensive. Devices with a FOM of 1,600 or above are of useful quality for security at sea.

Unfortunately, though, merchants rarely use this much more helpful FOM classification; instead, they still use the generation model. Generation 1 goggles claim to offer a 1,000-fold enhancement of the available light. Although this sounds impressive, most generation 1 devices produce too much visual noise to be of any real use in a piracy situation.

Generation 2 goggles are supposed to offer a 20,000-fold enhancement and the level of noise should be strongly reduced. Consequently, they are a more valuable tool for spotting targets in lowest-light conditions, although their quality is reflected in their price. The more advanced models' systems are frequently referred to as 'generation 2+'.

Generation 3 systems were constructed for even better enhancement (up to 45,000 times) as well as longer tube life. The technology usually also produces more noise and thus is considered only equally as useful as generation 2 or generation 2+ devices.

Although most higher-quality NVG are water resistant, it is worth checking their resilience in a marine environment. Some NVGs include a built-in infrared light source to brighten the image. Most of these light sources are rather weak, thus rendering them useless if employed at long ranges (>150 metres/yards).

On a pleasure yacht, monocular versions prove more practical when switching from device-assisted vision to normal sight. Due to their technology, NVGs project bright light into the operator's eye, more or less ruining their natural night vision for a while. Merely an annoyance when only one eye is affected, the loss of night vision in both eyes is much more serious since it renders the user temporary night-blind once he puts the device aside.

Thermal imaging equipment

Risk Level	III IV
Yacht situation	⚓ ≈
Security score	★★★☆☆

Thermal imaging devices are able to show the heat radiated from the objects on an internal display. This means that they work in total darkness since they do not rely on any other light source. They also work well in foggy and light-rain conditions.

There is a wide range of qualities on the market, ranging from lightweight hand-held devices with low resolution up to five-digit mast-mounted and gimballed models that combine thermal vision with zoom daylight cameras.

Thermal imaging devices are complex systems with many data points to describe their capabilities. When first considering whether or not to buy a unit, pleasure yacht skippers might first look at the target detection range. This figure is most important for security applications. To be of any help in rating the product, the type of targets needs to be stated. Depending on their size and typical heat radiation, targets strongly differ (think of an oil rig vs a person in the water). In this context, 'detection' means is that the operator will notice 'something is out there'. At the maximum range, it is not necessarily possible to recognise what exactly that thing is.

Detection range is dependent on many factors, the most important ones being the lens and the detector size (the more pixels the detector offers the better). To be of any use for detecting a pirate attack, the range for noticing a small boat should be at least 1 nautical

FIG 3.17: Two boats at night as viewed through a thermal camera – it also works in adverse weather. (Source: FLIR)

mile (NM). To achieve such a performance today, hand-held devices need a detector size of 640 x 512 pixels, and fixed modes of at least 320 x 256 pixels (due to better lenses).

Fixed models installed on the yacht are an alternative to the hand-held devices. Most products can be integrated into your existing network, thus displaying the images on your main display. They usually feature an extra control unit for panning, zooming, etc. Unless installed completely free standing, they will have annoying blind spots caused by parts of the superstructure. This is especially true for sailboats, which have both masts and sails in many potential installation sites.

Despite being useful in an anchorage, we feel that only the best (and consequently most expensive) fixed models are at least partially helpful in security situations on pleasure yachts at sea (where the watchmen have to constantly scan a 360-degree horizon for threats). Such high-end models claim to offer effective gyro-stabilisation and target following, both of which are necessary to keep a target on the screen on a rolling and yawing sailboat with a small crew. Nonetheless, these features do not solve the problem of actually detecting the target on the horizon in the first place.

Provided that the device is of sufficient quality (with a detection range for small boats that is greater than 1NM), a hand-held thermal imaging device used by a vigilant guard offers a substantial increase in security during dark nights and other adverse conditions while cruising high-risk waters. At present, offshore night-time attacks by pirates are very rare. Consequently, the usefulness of night vision equipment can be weighed against budget considerations.

Alarming and deterrence

VHF Radio, preferably with DSC

It is assumed that any seagoing cruising yacht will have a properly installed VHF radio on board: these are of great use in the anchorage as well as offshore.

By providing continuous, unmanned distress signalling at the push of a button, DSC/GMDSS (Digital Selective Calling/Global Maritime Distress Safety System) radios add an important feature to the arsenal. A DSC distress call creates a cacophony of beeping on all other DSC radios (ashore and on other boats) in range. This includes the boats of attackers who – if they operate such a radio – will also notice that help has been called for. At least one case of an attack at open sea reports that the pirates retreated after distress calls had been issued via VHF. DSC also transmits the boat's Maritime Mobile Service Identity (MMSI) number as well as the last known GPS position, thus supporting the vectoring of rescuers to the site of the emergency.

Even without DSC, VHF radios have proven helpful in anchorages where crews have coordinated an impromptu support net and left the radios tuned to a joint frequency during the night. These nets work both ways: during an attack in St Vincent and the Grenadines, for example, a crew who witnessed an attack on another boat continuously transmitted distress calls on channel 16. The relentless transmissions were heard on the vessel under attack and helped to unsettle the attackers. The crew of the attacked boat later reasoned that the assault was less severe and prolonged due to the constant broadcasting.

VHF is also used to hail, warn and eventually deter any suspicious vessels approaching, especially offshore. Warning vessels via VHF serves to avoid misunderstandings, by offensively ordering any approaching boats to turn away and letting them know that they will be met by fierce resistance if they don't obey.

When passing through closely guarded areas, such as the Gulf of Aden or the Strait of Malacca, VHF is the primary channel of communication to supporting military and coast guard units.

Hand-held VHF radios and walky-talkies

Risk Level	I	II	III	IV
Yacht situation			⚓	≈
Security score	★	★	★	☆ ☆

One waterproof hand-held VHF radio is often found on cruising yachts as a back-up for the main unit, as a way of communicating with 'shore parties' and for the moment when the crew has to step over to a liferaft. They use the same frequencies as a fixed main unit and usually transmit with either 1 or 5 watts. Some boats that do not have a remote control for the main unit on the helm also use the hand-held to listen to other vessels in the vicinity.

When anchored in seedy bays, a hand-held placed under the skipper's pillow is beneficial if the skipper needs to call for other boats or support from the shore without moving to the main cabin to reach for the fixed VHF radio.

Security-aware skippers planning to spend some time in Risk Level III or IV regions might invest in a second unit. This proves very handy when the person on watch plans to raise an alarm or warn or simply pass information to the crew below without leaving his station. Concealed back-up units can also be very valuable after a successful raid that has stripped the crew of any other means of communication.

HF radios with DSC

Risk Level	I	II	III	IV
Yacht situation	⚓	≈		
Security score	★☆☆☆			

High-frequency (HF) or single-sideband modulation (SSB) radios are great pieces of long-range communication equipment so long as they are configured properly. Such systems can easily make contact with strong coastal transmitters from more than 3,200 kilometres (2,000 miles) away. SSB DSC allows you to initiate a distress call using GPS position, just as its cousin the VHF DSC does. The difference is the long range. Coast guards and marine rescue coordination centres are still watching on several SSB DSC frequencies, but lower frequencies (2,181kHz voice distress and 2,187.5kHz DSC) are no longer monitored by many services.

While HF DSC is one good option when you need to call for help, general communication with HF radios is a bit harder as sound quality is not comparable with that of VHF even if both transmitter and receiver are well tuned. What's more, tuning and operating an HF unit while under stress in noisy conditions is a lot harder for the not-so-well-trained user.

Long-range HF transmission quality is dependent on the range, the sun's position, solar activity and weather conditions. It is still OK for calling a distress with DSC or verbally as you can hope that a distant MRCC will receive the call and relay it to the closest units to help you. However, to complete a MAYDAY call on a HF unit without DSC is very hard, especially during daytime hours when the sun in the sky impairs transmissions. A satellite phone would be the better choice for making long-range calls to known parties, day or night.

In some areas with a high-density security infrastructure (such as the Gulf of Aden), high-frequency communication is not specifically offered for summoning rescues. Instead, a regular call centre is standing by and can be contacted via a phone system. This fact may be an indication of the dwindling importance of high-frequency communication as a means of making distress calls.

FIG 3.18: *A VHF radio is important for a yacht's security. (Photo: Getty)*

You should bear this in mind since DSC-ready units that include the antenna, automatic tuning unit and ground/counterpoise are fairly expensive. The installation costs will vary with your type of boat and availability of capable technicians.

Satellite phones and messengers

Risk Level	I	II	III	IV
Yacht situation				≈
Security score	★★★☆☆			

Satellite phones work very much like mobile phones. However, in contrast to regular GSM (Global System for Mobile communication) mobile phones, they offer telephone connections in practically every spot in the world, including the middle of an ocean.

With the most relevant and capable marine rescue coordination centres on speed dial, skippers can use a satellite phone to call for help and hope for the fastest possible assistance available for their position. Usually, the nearest Maritime Rescue Coordination Centre (MRCC) is a good option, although it sometimes pays to call a MRCC that speaks English or the caller's native language. If the MRCC is not in range, it will relay the call with high priority to the centre closest to the site of emergency.

In some regions, supporting units have set up specific telephone numbers with call centres staffed 24/7 to assist any vessels in trouble. Such a telephone-based support infrastructure has worked very well in the Gulf of Aden: the coordinators know all vessels' positions, crew numbers and defensive capabilities. Consequently, they can quickly and effectively provide support it there's an attack or suspicious approach made on any civil vessel in the region. Without a satellite phone, crews will not be able to benefit from these services.

FIG 3.19: The Iridium Go access point. (Photo: Iridium)

The latest models of Iridium or Inmarsat phones also offer tracking (automatic uploading of GPS positions to a pre-defined address), data links, messaging and distress calling.

Satellite access points for data calls as well as satellite-based messenger systems are valid alternatives for the classic phone (which usually combines voice, data and messenger). Pure data access points such as the Iridium Go feature local WiFi and routing of mails through the satellite system. The bandwidth is sufficiently high that it can globally request and receive weather/GRIB data (General Regularly-distributed Information in Binary form, which provides reliable weather information that can be interpreted by weather and routing software on your on-board computer). It also features a distress button and GPS.

Messengers such as the SPOT or the DeLorme inReach offer a somewhat restricted set of features at a lower price. With these systems, you can be tracked, set off SOS alerts (including GPS position) and send SMS-style messages to similar devices or via email. They are able to receive short messages in return.

Many satellite systems (phone, data and messengers) are nowadays equipped with a distress button. This very powerful feature serves a similar purpose as the DSC distress function on an HF or VHF radio, adding global reach. Depending on the make and model, they will send a MAYDAY message to a rescue coordination service, which will then relay the message to a proper MRCC. Some of these devices feature GPS and will send the current position together with the distress call. Consequently, these systems can serve as a fully independent back-up to a vessel's Emergency Position Indicating Radio Beacon (EPIRB) or HF DSC distress call.

Especially if you are planning to use voice calls when feeling threatened, it is a good idea to invest in an external antenna for your satellite system. Most devices will not connect well (or at all) to their satellites while below deck or in the saloon of a catamaran. This lack of connectivity could force the caller outside while the boat might be under attack by criminals. The external antenna leading to a cradle inside allows the operator to remain relatively protected while still having the best possible reception.

FIG 3.20: DeLorme inReach, a popular satellite messenger. (Photo: DeLorme)

Some skippers argue that satellite phones are inferior to standard radios – such as VHF and HF – because a call won't be not heard by everyone in range who has their radio tuned to distress channels. While this is true, we think that the ability to quickly reach any professional MRCC in the world – where there's an experienced, official team relaying the calls with high priority – compensates for the limitations concerning 'public calls'.

One drawback is the monthly fees payable for many types of contracts. These have to be paid on top of the investment for the hardware. By contrast, HF radios have to be bought once and are free to use after that. When considering satellite communications, skippers should shop around for the technology (phone, data system, messenger) and contract type (pre-paid vs post paid) that best serve their needs.

In conclusion, we encourage skippers to seriously consider the use of satellite phones as a more modern means of communication, rather than a classic HF radio.

EPIRBs and PLBs

Risk Level	I	II	III	IV
Yacht situation	⚓	〰		
Security score	★★☆☆☆			

EPIRBs should be present on all vessels sailing offshore. They do an excellent job by alerting rescuers and helping to locate vessels or crew in liferafts after accidents and emergencies. Unfortunately, the process of actually receiving both the distress signal and pinpointing the position of a vessel in distress may take some time. Even when the device's position is transmitted (as is done by the slightly more expensive models that feature integrated GPS), the reaction to the call could take several hours.

EPIRBs are thus considered very helpful when coping with the aftermath of an attack. They would also be of some use in areas with a high density of potential rescuers (such as the Gulf of Aden in 2016) as they continuously send distress signals without any crew member having to oversee the process (as would be the case for a VHF call without DSC).

On the other hand, the delayed alerting process of EPIRBs would make a VHF call the primary channel for gaining the instant attention of supporters that are within a 25NM radius. What's more, since they also work silently, they do not directly unsettle any attackers in the same way that seeing red flares going up or receiving their target's MAYDAY calls on their radio would.

Personal Locator Beacons (PLBs) are a small version of the EPIRB. They use the same frequencies but feature smaller batteries, thus limiting their operating time. Some skippers hide PLBs in their lifejackets when travelling through Risk Level IV waters. The hope is that they will be able to take the jacket and PLB along if they are kidnapped, which would enable rescuers to track their position. To date, there have been no reports of a successful PLB-supported rescue.

Pyrotechnic distress signals

Risk Level	I	II	III	IV
Yacht situation	⚓	〰		
Security score	★★★★☆			

Emergency light and smoke signals are excellent at fulfilling their purpose, namely attracting attention and helping rescuers to find a vessel in distress. As the IMO COLREGs (the International Maritime Organization's International Regulations for Preventing Collisions at Sea) specify, any vessel venturing offshore needs to carry them for safety. The typical kit on every cruising vessel includes red hand flares, red parachute flares and orange smoke pods.

Pyrotechnics are very helpful before or during criminal attacks both at sea and in anchorages. Due to the fact that the meaning of these signals is common knowledge, any vessel deploying them when feeling threatened will most likely gain instant attention from other boats and the shore community. Criminals will think twice about sustaining their attack while such signals are being set off. There are also some reports of boarders having been successfully deterred by emergency signals in an anchorage and while travelling at sea.

As a method of last resort, crews could also use parachute flares as horizontally aimed warning shots or as an actual long-distance weapon. The SOLAS (Safety of Lives at Sea)-approved models found on yachts will shoot to a distance of 300m (1,000ft) and feature a burn time of about 40 seconds. If they hit a wooden or plastic boat without bouncing off, they are an extreme, lethal hazard for its crew; the magnesium flame will set fire to most flammable materials on board that it has contact with, including the hull itself.

FIG 3.21: *Parachute rocket flare. (Photo: COMET)*

Firing warning shots is fairly easy: simply point them in the general direction (a little higher) of an offensive vessel and trigger the flare. However, a deliberate aim is quite difficult to achieve without training. Even if it actually hits your target, the projectile has a high chance of simply bouncing off any surface without doing any harm. If a skipper intends to use parachute flares as a weapon, he should buy at least five for practice and some more to use as an actual weapon (without diminishing the stock kept for emergencies).

Shooting flares against a determined criminal will most likely not stop him from attacking you, however. Villains with burning magnesium rockets sticking out of their penetrated chest are a product of B-movie industry, not reality.

CAUTION: All models expel hot gases and material from the muzzle when they are set off. Some models also expel plastic parts and hot gases from the other end. If held incorrectly in front of the shooter's face or body or in the direction of other crew members, serious injuries can occur.

Flare guns

Risk Level	I II III **IV**
Yacht situation	⚓ 〰
Security score	★★★★★

Flare guns are the reusable cousin of standard emergency pyrotechnics. Compared to the latter's integrated design of ammunition and deployment platform, most flare guns are built as single-shot, breach-loaded pistols that can be used with a variety of ammunition to attract attention in an emergency situation.

FIG 3.22: *Flare gun.*

In Europe, 'Calibre 4' pistols with a fairly long barrel are used. They shoot single and multiple stars, which will rise to about 130 metres (yards) and burn for about eight seconds. Alternatively, ammunition that leaves a smoke trail for daylight use is offered. Calibre 4 pistols are compatible with red-and-white parachute rockets that will rise to 330 metres (yards) and burn for up to 30 seconds.

In many other regions, such as the USA, most guns are marketed in the 12-gauge shotgun calibre. Such pistols usually only shoot stars and streakers.

Compared to standard emergency signals, flare guns have some pros and cons. Some skippers like their ability to shoot projectiles without hot gases being expelled from the back (as in some models of parachute rockets). Consequently, there is added safety when they are fired through a hatch from a confined space below.

In extreme distress during passages of pirate-plagued waters, the situation may arise in which a crew wants to shoot flares over the bows of an offensively approaching vessel. Shots from flare guns can be aimed more easily, safely and precisely than those from standard parachute tubes, which enhances safety for both boats. Actually hitting an intended target still requires training with the devise.

On the other hand, signal pistols are neither quickly nor easily reloaded in an emergency (both naval or criminal). The operator is forced to use both hands for reloading and must have his spare shots stored in reach. This rather complex handling prevents him from holding on to the boat and enhances the risk of ammunition or the gun itself going overboard in rough seas.

The larger variety of available ammunition, however, means that flare guns can also be used for issuing standard emergency signals, such as warning other mariners in near-collision situations (white flash-bangs are available for this).

There are some ideas about the usability of flare guns for self-defence. Although it is acknowledged that less-determined offenders will be intimidated when faced with a flare gun, it will most likely not impress a more aggressive criminal.

Various reports have been analysed to cast some light on their true value as weapons that cause injuries. The results support the notion that they are of limited use in a confrontation: the persons shot at were perfectly able to walk and even fight, despite having being hit in vital body regions. There are two reliable reports of humans shot from a US-style 12-gauge flare gun using standard flare ammunition. One of these states that the attacker hit by the flare 'laughed and carried on with the attack'. The second covers an incident in which a suicidal person shot himself three times in the head. This man was in a critical condition when rescued, although he must have been able to load and fire after two point-blank shots, and wasn't dead.

Adding to the lack of physical effect, the one-shot capability is a problem in conflicts that involve more than one attacker. There is also a high risk of projectiles bouncing off the target (either a boat or a person). Such ricochets could trail off into the water or start a fire on the attacker's or defender's boat.

We can, therefore, conclude that flare guns are a very useful tool for signalling and, in a criminal emergency, firing warning shots. When it comes to an actual confrontation, though, more reliable and multi-shot weapons such as a pepper spray are most likely the better choice.

Flare guns are regulated in many countries and some states of the USA. When entering a country, skippers need to check whether the obligatory report on firearms includes emergency flare guns.

Signalling lasers

Risk Level	I	II	III	IV
Yacht situation				≈
Security score	★☆☆☆☆			

In the last couple of years the signalling laser has gained a good reputation in terms of serving similar purposes as the hand-held magnesium flare. The beam emitted is bar-shaped, which makes it easier for a possible rescuer to spot it if it's moved along the horizon. These models feature rather weak laser sources to avoid injuring the parties that you call for help. For this reason, although such lasers are a good back-up or substitute in a general naval emergency, it is not recommended that you rely on them to signal for help while an attack is ongoing. Their deterrent potential is regarded as being low and the operator would have to constantly scan a target sector to indicate his emergency while the attack is underway – parachute rockets or smoke that keeps burning unattended once deployed are a much better option.

Acoustic systems

Risk Level	I	II	III	IV
Yacht situation		⚓	≈	
Security score	★★★★☆			

Although so-called 'long-range acoustic devices' (LRAD) are sometimes used on commercial ships, most yachts lack either the space or the power supply to make them part of their defensive arsenal; instead, skippers should consider installing a powerful foghorn and hailing device on their boat if possible.

FIG 3.23: Electric horn and hailing device. (Photo: Kahlenberg)

FIG 3.24: Compressed-air model. (Photo: Kahlenberg)

Foghorns and sirens will not dazzle an assailant as an LRAD would, but they do alert crew members, boats in the vicinity and the onshore community. At the same time, they signal to attackers that they have been spotted and may unsettle them, since detection by law enforcement or other supportive groups will probably be imminent.

Such devices can be controlled manually from the helm or navigation table and some VHF radios include a fog-horn control. These can be set to continuously sound the horn at programmed intervals and sequences.

Electric and compressor-actuated variants are available in a wide range of prices. The electric version is less complex and easier to install, as it only needs an electricity supply. Some of these models also work as a hailing device by amplifying the communication from a VHF microphone.

Models working with compressed air that are suitable for cruising yachts feature an integrated compressor, whereas the larger models would need a hose supplying pressurised air from an external compressor (which are rarely available on smaller vessels).

Most produce a signal strength of 100–130dB at a 1-metre (yard) distance. For security, stronger ones are more effective since their range and general potential to raise help is greater.

Alternatively, simple breath-actuated horns as well as models that work with a can of compressed air could be used to some effect. The drawback of such 'manual' devices is that the operator is occupied with signalling and is therefore unable to engage in other useful activities during a criminal attack.

Deck lights

A bright deck light is an effective deterrent in many situations, especially for boats that are not anchored alone or that are anchored close to a lively beach. Lights signal that someone is home. They also increase the real and perceived risk of being detected (by people inside as well as by third parties on other boats or ashore), thus effectively reducing the element of surprise and initiative for the assailant. Bright lights also put any boarder in an unfavourable situation, since he will be very visible on deck, unlike the crew in the dark rooms inside.

Unfortunately, the deck lights on many cruising yachts have two drawbacks: first, quite a few still use energy-hungry halogen bulbs, placing a considerable strain on a yacht's batteries when switched on during dark hours; and second, the lights are installed on the frontal face of the mast (often combined with the steaming light). This configuration leaves the most vulnerable parts of the boat – the transom, cockpit and companionway – in the dark.

One simple solution is to exchange your halogen bulb for a bright LED version. This action will probably allow you to leave it on during the night without adding extra power to your system. It might also be a good idea to install rechargeable LED floodlights in a position that illuminates the sensitive areas of your boat. Some monohull skippers have simply hung such lights from their boom. Cautious catamaran skippers put a strong lantern on the cockpit table or point a heavy-duty spotlight from the inside of the saloon to the stern section of their cat. This configuration can even be hooked to the boat's battery, ensuring all-night illumination.

Replica firearms

Risk Level	IV
Yacht situation	≈
Security score	★☆☆☆☆

Replicas can be bought from specialist retailers or built by the crew. They are cheap and legal in many countries, even though so-called airsoft/softair models look stunningly like real guns. They are helpful to warn or intimidate the less-determined and less-well-armed offender. Naturally, they are useless in a firefight.

The airsoft versions are usually made to fit the dimensions of the real thing. Some are constructed from a metal/plastic combination that makes both their looks and handling truly realistic to the observer. Airsoft replicas usually fire electrically propelled 6mm plastic bullets that are of no use in any confrontation. Their price will vary depending on their quality and brand. Although the market offers special-forces models of rare guns featuring tactical lights, laser sights, etc, it is not wise to invest in such products. It would simply be very unlikely that either a skipper or a security team on board would use such weapons. It is better stick to the serious but simple AK-47, Colt M4 or Heckler & Koch G3.

FIG 3.25: *Airsoft replica of an AK-47. (Photo: Berswordt)*

Depending on the country in which the vessel is currently residing, replicas might be difficult to obtain. However, since they are legal and not considered firearms in most places, it should be safe to take them on board while in dock. It is up to you to check the rules and regulations for the country you are in.

And they are worth having, especially if a Risk Level IV passage is planned – even a well-made DIY-replica can prove useful in an offshore situation. Since early deterrence starts when an attacker is about 500 metres (yards) away, a replica could fool less-experienced or less-determined attackers. For this reason, some skippers who oppose real firearms opt to carry them.

For others, replicas may be added to the arsenal in combination with real guns, thus giving the impression that the whole crew is armed. In this instance, should mere presentation of the guns fail to deter the suspicious vessel, warning shots from the real gun would make attackers think that all the weapons are real.

FIG 3.26: DIY replica made from a leeboard, a defunct antenna, an old belt and a high-power laser. (Photo: Berswordt)

Replicas work best when they are used in combination with other tricks, such as clothing and general crew demeanor that makes you appear to be a professional security team (see 'Sailing in pirate-plagued waters', page 130). We have actually used such tactics in the Gulf of Aden to keep suspicious vessels successfully away from our convoy.

Replicas lose all value the moment any serious attacker with a real gun fires a shot. For this reason, they are not recommended while at anchor or in any other situation where a first shot is imminent early in the confrontation – such as with a nightly boarder in a remote bay.

Alarm systems

Alarm systems aim at detecting an intruder and alerting the crew of a boat as well as bystanders ashore before damage is done. The system's alerting mechanisms usually unsettle and – in the best cases – chase away the attacker. When comparing the costs and effect, alarm systems are one of the most efficient and effective pieces of equipment. It is highly recommended that every yacht is fitted with at least one compact alarm system.

FIG 3.27: *Compact alarm system. (Photo: Amazon)*

The costs for compact devices range from very little indeed for a simple product up to many thousands for networked systems.

Compact products are battery operated and combine detection capabilities with an alerting method in one device. Even the cheapest ones are bundled with a remote control, allowing the crew to deactivate a system that has monitored the boat while the crew was ashore. Short message service (SMS) features are included in some.

Most skippers position systems so that they monitor the expected boarding sites as well as the most reasonable entry points to the yacht's interior (see pages 50–4, 'Ship layouts and security').

Models with a flat underside do not have to be permanently installed using wall or ceiling mounts; instead, they can simply be placed wherever is most appropriate. This works fine on most plane surfaces, whereas an improvisation has to be made for rounded or slanted parts of your boat or when strong winds are blowing and threaten to topple the system.

Some of the cheapest ones may not be watertight and would fail in strong rain. The skipper should, therefore, ensure that the product features at least an IP 2, preferably an IP 3, rating for water resistance; otherwise, the devices have to be positioned indoors during rain, thus limiting their usefulness. Due to the very low price of these models, some skippers have four or more on board and simply replace damaged or unreliable ones as they fail.

Systems that are more complex separate detection (i.e. sensors) from alerting parts (i.e. sirens and lights) and network them via a wired or wireless connection. They often employ a central base station that adds some extra features, such as a silent alert, panic button (a device that the crew can use to manually trigger the alerting siren) and the integration of SMS messaging to a mobile phone. The more elaborate systems also offer indoor and outdoor cameras as well as video transfer to a remote device.

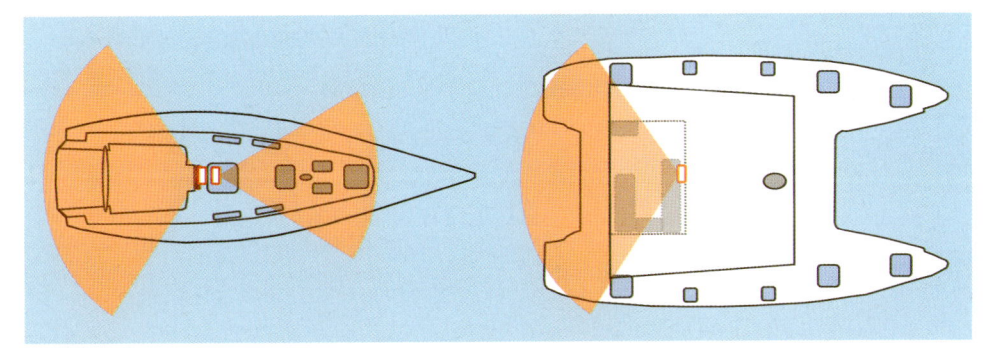

FIG 3.28: *Placement options for compact alarm systems on monohulls and catamarans.*

Their disadvantage is that they normally have to be permanently installed somewhere. Especially on small boats, having an ever-present alarm system is not what some skippers want or like. Such devices also use on-board electricity to a varying degree, so their energy consumption needs to be regarded when planning the energy balance for a boat. If a part or all of the system is damaged, they are usually more difficult to replace in remote parts of the world.

Most systems (both cheap and expensive) use infrared sensors to detect an intruder. These work fine in an open-view environment, although they will not work through GRP walls. Some cannot even penetrate the glass panel of a cruising catamaran's saloon window. For this reason, crews must find places that allow for a good coverage outside. This is quite a challenge, given that some sensors of permanently installed systems are constantly exposed to the harsh marine environment. Not many products will work for a longer time in such conditions.

Some products (both compact and more elaborate) use radar for detection. Radar is capable of penetrating glass and GRP, giving crews the option to place the device indoors, thus protecting it from weather and – rather unlikely – from being tampered with by the intruder. The downside of compact systems with a radar detector placed indoors is that the alarm is not heard well outdoors, thus limiting its ability to alert third parties and startle or deter the offender. When used by networked systems that trigger outside sirens, however, this is not an issue.

To alert crew or third parties, most simple systems rely on a very loud electronic siren. The better ones will emit sounds louder than 100dB – some claim to emit noise up to 130dB. This will wake even the sleepiest skipper and alert others in the vicinity. More elaborate versions feature their own sirens but could also be hooked to the foghorn, on-board floodlights and an SMS alert system, warning the crew that is ashore of an intruder on board.

Cameras and CCTV surveillance

Risk Level	II III IV
Yacht situation	🛢 ⚓
Security score	★☆☆☆☆

Cameras can serve multiple purposes: their mere existence (if positioned conspicuously) will deter some types of attackers, they can help crews to evaluate an ongoing attack, and they support law enforcement after an attack.

In domestic security, deterrence is a perceived benefit of bulky camera and illumination systems. They are installed very visibly on buildings' corners as well as relevant gates and passages. Unfortunately, the average cruising yacht does not offer many places where such a system can be installed. Additionally, recent research on the effect of domestic CCTV systems on burglary and robbery shows that they made no differences whatsoever.

Some brands offer fixed camera/thermal systems that are installed on the spreaders or the gantry. Since these are not very common and many thugs might not be able to identify the dome-shaped devices as cameras, they are thought to have limited deterrence potential in more remote regions.

When a crew suspects that their vessel is about to be boarded in an anchorage or at dockside, a camera could be helpful for identifying the type and severity of the threat from a safe position below decks. How many people are boarding? What is their armament? What is their position and where are they heading? If a fixed camera is installed in a position that allows for deck surveillance (some only point at the sea ahead), they might be suitable for such a task. However, when push comes to shove on a small boat, time is a crucial element. Gathering in front of the screen of a CCTV system, let alone panning and zooming to get the correct angle for identifying any boarders, is most likely not time well spent because the offenders will move fast towards vital areas such as entrances and hatches.

FIG 3.29: CCTV cameras can be used to view sections of yacht and sourroundings. (Photo: Raymarine)

Surveillance can also be improvised by using modern action cameras, which are common on cruising boats today. The more sophisticated models feature a WiFi connection to a tablet, smartphone or notebook. The users can control basic camera features and see the camera's video feed on their device in their cabin. Most models come with a wide-angle or fisheye objective lens, which allows greater flexibility when placing the device. If such a camera is set behind a catamaran's sliding window or hidden between some lines close to the main hatch, it could be helpful in determining the crew's reaction toward boarders outside.

Camera footage might also be used after a robbery or burglary if it was recorded on a device that was not snatched by the offenders. Unfortunately, direct upload to the 'Cloud' is usually not an option at remote anchorages and even recording to a hidden storage device is a feature rarely offered on any but the most expensive models.

It can only be guessed whether the local police are capable of or interested in using camera footage to identify offenders after an attack.

Warning signs

Risk Level	II III IV
Yacht situation	🛢 ⚓
Security score	★★☆☆☆

Warning signs are very simple to produce and contribute to the overall security of a vessel at anchor or in a marina. If placed at the most likely boarding points, they can deter the less-motivated or more timid burglar and thief.

The signs could warn of CCTV cameras, alarm systems, a dog or security team on board (depending on the situation and risk level in which the yacht is travelling). Since offenders often have a chance to spy out the yacht during the day, it is wise not to announce anything that is obviously not there.

Some organisations supporting security missions in pirate hotspots such as the Horn of Africa recommend installing warning signs in English and the local language to announce the presence of electric fences. Given that such defences are mostly not feasible for the average cruising yacht (and other barriers are conspicuous in themselves), it is not recommended here.

The more unmistakable the signs or icons used on the sign, the better it will work around the world. English is usually the most widely understood language.

High-powered flashlights

Risk Level	I	II	III	IV
Yacht situation	🛢	⚓	≈	
Security score	★★★★☆			

Flashlights have come a long way since the old days of filament lamps casting a short-ranged and short-lived yellow light onto objects of interest. Modern hand-held torches feature multiple high-powered Cree LEDs and are capable of casting a beam stronger than the headlights of today's cars. They are hybrid tools when it comes to security. On the one hand, they help you to identify any suspicious activity – a vessel in the distance, a skiff loitering in the anchorage or a stranger on your deck – and on the other hand they can be used to dazzle, blind and disorient nightly boarders, helping defenders to gain and keep the initiative in a possible confrontation.

The strongest versions are often specialised. Throwers cast a narrow 2,600-lumen beam to a maximum distance of a little more than 1,000 metres (yards). Flooders are optimised to light up the environment in a wide angle in front of the user. The top models are tagged with more than 7,000 lumens and will raise the day in front of you. However, their effective range is only slightly more than 60 metres (yards). There are some hybrids that attempt to combine the features of both throwers and flooders with varying success.

Most models feature several modes in order to provide less brightness and prevent the device from getting too hot to hold. A strobe mode is often integrated and this could be used for raising attention on other boats or the shore.

FIG 3.30: *Beamshots of a modern thrower and flooder. (Photo: www.traumflieger.de)*

Quality models are not prohibitively expensive, but bear in mind that the item may not come with the necessary high-quality lithium-ion rechargeable batteries and a charger, so this would need to be bought separately.

Some models on the market show off with higher lumens. These so called 'china lumens' need to be divided by three to obtain a realistic value for such flashlights. Models from AceBeam, ThruNite and Fenix were correctly labelled and had very good reviews from aficionados at the time of writing.

If you have to decide which model to buy, try to shop for a high-quality hybrid or a thrower because they are a little more versatile: you can use them to identify suspicious activity over long distances and effectively (and temporarily) blind uninvited visitors, and the narrow beam is not much of a problem on small or medium cruising yachts.

Defence and resistance

Propeller-fouling devices (PFD)

Risk Level	III IV
Yacht situation	≈
Security score	★★☆☆☆

Propeller-fouling devices are used against boats pursuing your vessel in its wake. They work by entangling and thus blocking the propeller, allowing the attacked vessel to escape to safety while the criminals spend time freeing their prop. Consequently, PFDs are only useful while travelling at sea and against targets approaching from behind.

There are recent reports from at least two crews who managed to successfully disable the propellers of suspiciously approaching boats.

At present, PFDs are not marketed, although skippers are quite creative in improvising

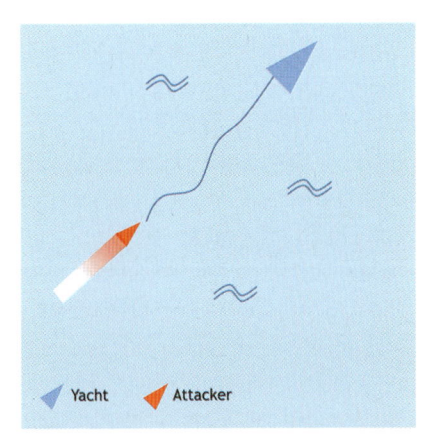

FIG 3.31: Mode of action of propeller-fouling devices.

them. Two designs have been reported. The simpler one is made from a swimming line that is fixed on the transom dragged behind the boat. This could be made from cheap polypropylene material or much more expensive high-quality polyethylene (marketed as Dyneema™ or Spectra™), since this latter would most likely remain on the surface when towed. Either way, the line's diameter could be anything between 2mm and 4mm. Dyneema™ lines could be even thinner (think about those high-strength braided fishing lines). Wider diameters than this would make the line more conspicuous in the water as well as more complicated to store at your transom, so are best avoided.

Most skippers prefer dark colours, such as brown, black or navy blue to camouflage the line in the water. The length should be at least 100m (325ft). It might be a good idea to simply rig a free-spinning reel or a 'quick-release bunch' with a couple of hundred metres of line on your transom's live lines, which could be quickly deployed in times of trouble.

More elaborate designs include a net at the end of the lines towed. Some even attach simple trawl doors to keep the net spread. Currently, we are not aware of any reports that compare and rank the net design vs the simple line. For this reason, we suggest you go with the simpler line version. If skippers wish to cover a wider portion of their wake with fouling gear, they might consider rigging two or more lines.

The PFD design must include a more-or-less automatic release or a pre-determined breaking point in case an adversary's propeller successfully catches it; otherwise, damage to the transom may occur or the attacked vessel could end up towing the pursuing vessel on a particularly strong Dyneema™ line.

PFDs limit the manoeuvrability of the towing vessel since tight turns could lead a ship to cross its own PFD and consequently disable its own prop.

Bear in mind that you will most likely be very agitated when it is time to deploy your PFD, so it should be very easy to get it properly overboard without the need for much time or thought.

Blocking devices and barriers

These are all devices that will delay the boarding of any uninvited intruder on your boat. They range from your standard lifelines to concertina wire barriers. As always, their usefulness greatly depends on where a vessel is and what the crew expects.

To be effective, barriers have to be at least semi-fixed and thus impair the crew's mobility having quite a negative effect on how much you can enjoy a stay at an anchorage.

In most regions that pleasure yachts visit (Risk Levels I–III), extra barriers are not necessary for safety. Nevertheless, skippers may think about closing their lifelines at the transom each night, making it less likely that the occasional burglar or robber will board unnoticed. If a place is so dangerous that a skipper has to consider using barriers, then it is probably best avoided altogether. When stranded in truly gnarly anchorages or docks, a 24-hour watch is the better alternative for the boat's safety.

Nonetheless, barriers do add to the crew's peace of mind and safety when at sea in areas with a risk of pirate attacks because they make the boat appear more fortified, and therefore less of an easy target. Many opportunistic attackers might think twice about an attack when seeing barricaded transoms. A good barrier can also at least slow down a boarding or else funnel boarders to places that favour the defenders if there is a conflict. However, it will most likely not keep determined professional pirates or terrorists from boarding your boat. These criminals are willing to take on commercial ships with a freeboard higher than anything that a pleasure yacht has to offer.

Improvised barriers

Risk Level	IV
Yacht situation	〰️
Security score	★☆☆☆☆

Any construction that is capable of slowing down a boarder's progress from his vessel to the deck of the defenders can be helpful during an offshore pirate attack.

Although attackers from large vessels with high freeboard will be able to board from any direction they please (as they only have to step down to the defender's deck), it is more common for offenders in skiffs to carry out attacks. Unless they are athletic enough to pull themselves up over the gunwales, they will try to enter via the transom or the bathing platform.

Consequently, blocking access to the stern can give the defenders a slight tactical advantage. To improvise a barrier, such simple things as fenders can be fixed on the platform itself and the stairs leading up. A set of fenders tied to form a platform of rounded surfaces (without sufficient room to step between the units) is not easily negotiated on a vessel moving in the waves. Tied across steps, fenders can also be used to make stairs a little more difficult to climb. Another design employs planks or plywood panels that are fixed over stairs and the platform to form a slanted plane, which is also difficult to climb.

As previously mentioned, such barriers will not keep a determined offender from entering the boat, although they might discourage opportunistic ones. Moreover, professionals could be forced to use their hands to climb barriers, rather than simply stepping over to the attacked vessel with their guns aimed at the crew. Those moments when the attackers' hands are busy could allow for some last-resort defensive moves by the defenders.

Barbed wire

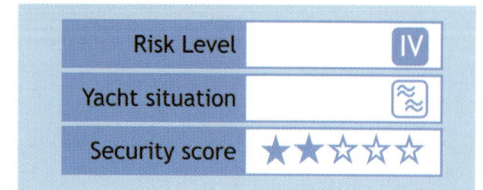

Risk Level	IV
Yacht situation	≈
Security score	★★☆☆☆

The use of either type of barbed wire described below as a barrier has to be carefully considered by the skipper. On the one hand, it can effectively impair the ability of boarders to come on board. On the other hand, there are some downsides: for instance, barbed wire is difficult to install without damaging the boat, meaning that crews have to be very careful when fixing it to the vessel. Wherever the barbs touch a vessel's surface directly, they will damage the gelcoat or varnish of your well cared for boat.

The barriers are also a risk to the crew on a boat constantly in motion, especially on smaller boats, where a foot set wrongly could send a crew member into razor wire. A person's degree of injury will worsen with every wave rocking the hull, and bigger waves pose a bigger threat. Coils of barbed wire are bulky and hard to store. Consequently, it is unfeasible on most yachts to have them contantly available to be installed at seedy anchorages (even ignoring the unpleasantness of residing in a vessel that looks and feels like a prison hulk).

Depending on the boat type, barbed wire cannot be applied to all sections of the vessel. Motor yachts theoretically allow for an all-around barrier, whereas non-heeling catamarans allow for barriers aft of the room that is needed for any larger headsails; installing it too far to the front would damage a gennaker and other sails that are guided outside the lifelines. On monohulls, the wire is limited to the transom or bathing platform as most of the sides can be submerged in the water, which would tear most wire barriers off the boat.

FIG 3.32: Barrier made from loops of razor wire around regular barbed wire. (Photo: H A Schwetz)

FIG 3.33: Razor wire is very effective at deterring boarders but is hard to store and dangerous to crew if not handled correctly. (Photo: Creative Commons)

There are two general types of barbed wire: the classical version that we know from farmland fences and a type called concertina wire, which is often used in military or other higher-level security applications.

Regular barbed wire is the less-effective barrier as it features fewer barbs with longer distances between them on the wire. Attackers can therefore grab barb-free sections more easily, even with unprotected hands. What's more, each barb is less effective at clinging to objects or injuring intruders. However, it is more easily handled by the crew during installation and the cruise itself and it will also be more forgiving if crew members hit the barrier by accident. It will needed to be replaced fairly frequently, since unless rarely marketed stainless-steel wire is used, the barriers will quickly rust in a marine environment. Barbed wire of non-stainless-steel quality is available in most hardware stores around the world.

Cincertina razor wire features more, sharper and longer barbs on the same length of wire. It is thus much more effective as a barrier, but also a lot more dangerous when handled by the crew or if someone falls into the barrier by accident. Depending on the country you are in, shopping for razor wire can prove tedious since it is not as widely available as regular barbed wire.

If you are still considering using barbed wire, then it is important to note that there are currently no reports of attacks in which its use made a difference. Barriers made from it are most likely unable to deter professional criminals with long-range weapons.

Electric fences

Despite sometimes being discussed on internet forums, reliable electric fences are not easily installed on a pleasure boat (if at all). Insulation is the greatest challenge on low-freeboard vessels when sailing and heeling. Generally, the systems are too complex and expensive to be of any use to yacht crews.

Firearms and other weapons

Firearms are the most effective and yet the most controversial security items on board any pleasure yacht and you should ponder long and hard about whether they are required, legal or advisable.

First, think about the general idea of firearms: do you really want to spend time in regions where you might actually have to use such weapons for self-defence? Unless you are an extreme danger-seeker, you probably would not enjoy the trip in the first place. If you stick to the routing strategies described in the 'Planning Your Cruise in Uncertain Regions' section (see pages 36–47), you are very unlikely to need a gun for self-defence.

Circumnavigators or explorers who choose or need to travel through pirate-plagued waters or near any Risk Level IV area would add a considerable extra weight on the security side of their security balance by carrying firearms. Skippers planning such routes have to decide whether they want to buy that extra safety on this part of the trip at the cost of increased bureaucracy involved with declaring your weapons.

In many nations, firearms are tightly controlled or illegal. Depending on a vessel's flag, it may thus be illegal for a skipper to keep them on board. This does not apply to vessels under the jurisdiction of countries that allow the possession of firearms. All others would risk criminal investigation if they ever had to face justice in their country after having used their guns in international waters. Consequently, before taking a gun on board, check the laws that apply in the country in which your boat is registered. As a general rule of thumb you could assume that if you need a permit at home, you will also need it on your vessel. If you are obliged to store the guns in specific containers at home, the same applies on board.

Bringing a firearm into any country legally is a major hassle as it always leads to longer check-in and check-out processes. At present, no known cruising destination allows a vessel to enter their waters without declaring the firearms on board. In some countries, crossing the border without prior approval is illegal in itself and you will face criminal charges if you attempt to declare them without having obtained that permission before entering their waters.

Some countries will take the declared guns into customs or police custody in the port of arrival. This is a major nuisance if you plan to exit from a different port. In the best scenario, you will be allowed to keep your weapons on board locked in a safe container, which is then sealed by customs.

As a result, the legally declared gun will be in custody or a locked and sealed container in most countries you visit. Fetching a gun from the sealed container in self-defence will take some time. Actually shooting someone (or even presenting a broken seal while checking out) will most likely lead to some serious questions and possibly prosecution by local law enforcement.

So, what are the benefits? Firearms are very effective in both deterring possible offenders and defending yourself if there is a conflict. They easily transform a fragile senior skipper into a very dangerous opponent for criminals. Merely presenting a gun to any attacker not armed with guns himself will most likely lead to the abortion of the attack. Warning shots will even scare away the most persistent offenders. Finally, a firearm would give you the option to attempt to fend off an attacker who is carrying a gun himself by taking a risk and 'shooting it out' – a situation that is most likely to arise if skippers ignore the routing recommendations of this book and travel to Risk Level IV areas.

However, engaging in firefights with terrorists or professional pirates is suicidal if your crew is not thoroughly trained in the use of their weapons as well as combat tactics. Unless you plan to rather die than be captured, never start a gunfight without having the initiative. It can and does go wrong, as can be seen in the tragic case of the German crew trying to fend off Filipino terrorists in November 2016, which left one of the yacht's crew dead and the other captured. The crew member, who was killed on the spot, was reported as having engaged at a group of eight to ten terrorists armed with automatic rifles.

If you choose to take a firearm on board, also take a lot of ammunition and plan its storage in dry conditions. To effectively use your gun, you must plan to train with it at least once a month with a minimum of 20 shots per session (unless you are a very experienced shooter before you set sail). Do not expect to be able to legally restock while sailing as the laws in most countries will not allow you to buy ammunition.

The choice of firearm should be determined by its intended use and by the ability of the shooter. To simplify a discussion that spreads over thousands of threads on the internet, it is recommended that you judge the weapons by their ability to deter upon sight (which is a high priority), and their effectiveness in mid-range combat and close-quarter battle.

Handguns: revolvers and pistols

Risk Level	III IV
Yacht situation	⚓ ≈
Security score	★★★☆☆

Handguns are, compared to rifles and shotguns, less effective in deterrence because they are not very conspicuous. Unless you fire warning shots, attackers simply might not clearly see and be put off by your weapon. Assailants might also know their restrictions concerning range, in which case serious pirates with rifles will most likely start to effectively engage you from a distance at which you are no threat to them.

Due to their size, though, handguns excel in close-quarter combat in confined spaces. In simplified terms, one could say that a handgun is great if you have to use it while the attacker is already on board or even indoors. However, this is a situation that should not arise unless you ignore most other advice in this book. We feel that handguns are inferior to other firearms as they create a high risk of skippers being put into a situation in which they actually have to shoot and seriously injure someone late in the sequence of an attack due to their lower deterrence potential and shorter range. Since this would have to be done at fairly short ranges, the shooter would be in great danger himself, especially if the attacker is equally armed.

Owing to a handgun's reduced potential as a deterrent, the bearer needs to ensure that he is well trained regarding its use (more so than for other types of firearms, especially the shotgun). This includes quick readying of the gun as well as actually hitting moving, belligerent targets while under stress. Only good, stationary shooters hit torso-sized, immobile targets 20 metres (yards) away with a handgun. If not continuously training, shooters would be wise not to assume that they can hit a moving human that is further away than 5m (16ft) while under stress. Consequently, only the most experienced shooters should consider a handgun on their vessel.

When a skipper chooses to buy a handgun despite its limitations, a modern pistol of high quality carrying a high-capacity magazine is a good choice. Such guns give you an advantage over revolvers as they carry more rounds, which allows for a few missed shots. One advantage of the revolver might be its simple, rather fail-safe design. Then again, a modern and well-maintained pistol will most likely not fail in an emergency.

The preferred round is a 9mm Luger with expanding bullets (hollow point, jacketed hollow point or the like). This calibre ammunition offers a good balance between the number of rounds in the magazine, the recoil, the damage to a typical human target and the risk of accidentally injuring uninvolved parties, given that such projectiles do not penetrate bodies and walls as well as full metal jacketed versions.

Rifles

Risk Level	III IV
Yacht situation	⚓ ≈
Security score	★★★★☆

Rifles are difficult for a would-be-attacker to overlook when presented. Most types appear sufficiently impressive that you can make your point even over a longer distance or in lower-light conditions. Unless you are packing a front-loading musket or BB gun, their potential to deter an assailant and thus avoid a confrontation altogether is much higher than that of handguns. If you use one to fire warning shots, both the flash and the bang significantly exceed those of a typical handgun. Rifles shoot regular bullets and need to be well aimed.

In a situation when an actual confrontation is inevitable, rifles have much greater effective ranges than handguns: 20 metres (yards) for most shooters, 100 metres (yards) and above for expert shooters in calm seas and air. This adds to their deterrent potential because more professional attackers will know this. That said, most rifles have quite long barrels, which makes them difficult to handle below decks. This could be a disadvantage when using them inside or when getting them ready and while manoeuvring them outside.

If firearms need to be taken on board, rifles are the best choice for more experienced shooters (as compared to shotguns). The optimal rifle would be a rather short-barrelled, semi-automatic hunting rifle with a five-or ten-round magazine in a small calibre (such as a .223 Remington) or medium calibre (such as a .308 Winchester).

A hunting rifle is said to cause less trouble with host countries' customs officers compared with military-looking weapons. High-quality rifles will also be able to cope fairly well with naval conditions if maintained properly, and short-barrelled rifles are easier to handle below and above decks. They have a little less power and range, although this is mostly irrelevant in moving, windy conditions. The shorter barrel usually produces a larger, more impressive muzzle flash (good for deterrence, bad for your night vision when fired in the dark). Compared to manually operated bolt-action or lever-action models, semi-automatic rifles are able to shoot faster and allow the shooter to 'stay on target' when firing multiple shots. A downside is their higher complexity and higher level of necessary maintenance.

Larger calibres, such as the .308 Winchester, offer a flat trajectory and good performance at ranges up to 150 metres (yards) – without much compensation for bullet rise or drop. They will also penetrate most materials of an attacker's boat, giving the crew the option to engage targets who are concealed by parts of their vessel. Smaller calibres such as the .223 Remington are much easier to shoot due to a weaker recoil and lower overall weight.

However, they feature a little less power to penetrate boat sides should an open-sea shootout with pirates occur. They could be a good option for the less-experienced or less-powerful rifle shooter.

Any full metal jacket bullet from the recommended calibres will seriously harm and, if it hits the centre of his body, most likely kill any attacker. They are cheap and, if aimed at the extremities, leave some room for disabling an opponent rather than inevitably killing them – something that is not an option with expanding hunting bullets that cause terrible, difficult-to-treat wounds.

Shotguns

Risk Level	III IV
Yacht situation	⚓ ≈
Security score	★★★★☆

The third type of firearms is shotguns. Most models typically come with large-calibre barrels and medium-size magazines, giving them an impressive deterrent potential. There are many stories of home defenders reporting that the mere sound caused by the pump action when loading a shell from the magazine to the chamber scared away intruders, thus avoiding a confrontation without them having to fire a single shot. Shotguns are loud and cause very conspicuous muzzle flashes when fired. Consequently, warning shots are impressive demonstrations of force and determination.

If a confrontation cannot be avoided, shotguns are effective in short to medium ranges up to 30 metres (yards). Due to the various ammunition types available, they are versatile in a conflict situation. Beyond 30–40 metres (yards), they lose both accuracy and effectiveness.

Shotguns are consequently the preferred all-around weapon for rather inexperienced shooters if a firearm needs to be on board. Given that it is often people who are not very familiar with guns who take a shotgun with them, a large-calibre (12-gauge), shorter-barrel pump-action version with a normal shaft is recommended. If weaker crew members also need to be able to use the gun, a smaller-calibre (20-gauge) could be considered. The manually-operated pump-action version is simple in design and compared to semi-automatic models is a little more resistant to the naval environment.

Shooters experienced and comfortable with servicing their weapon regularly will most likely choose the more efficient semi-automatic variant. The shorter barrel eases manoeuvring below decks and in confined spaces and the normal shaft – rather than a pistol grip – allows for better aim at medium ranges, less recoil and its less-military appearance makes the clearance process at customs easier.

Magazine capacities usually vary from five to nine shells. Since pump-action magazines are aligned as tubes running parallel to the barrel, very large magazines lead to longer guns. Skippers have to make a decision concerning whether they want more magazine capacity or shorter barrels for better manoeuvrability.

There are two very different ammunition types available for shotguns: slugs and shot. Slugs are projectiles in the diameter of the barrel. They easily penetrate boat-side material and cause mayhem when they hit a living target. They travel further than shot but are more difficult to aim as they comprise only a single projectile. By contrast, shot is made of many balls, each smaller than the barrel's diameter. These projectiles disperse the moment they leave the barrel, thus covering a larger area and making aiming easier. Shot varies in size and this affects the number of projectiles packed into one shell. The most frequently used version for self-defence is the so-called '12 gauge 00 buck shot' ('double-ought buck shot'). Most shells of this type will fire nine lead balls per shot from a 12-gauge shotgun.

A word about range: the mentioned 00 buck shot will disperse the nine pellets between 20 and 40cm (8–16in) over a range of 20 metres (yards). The spread will be even larger when the range is 45 metres (yards): 80cm (32in) will not be an exception. Consequently, you have to assume that most of your projectiles will miss if they are shot from distances further than 25 metres (yards) away.

Laser pointers and high-power lasers

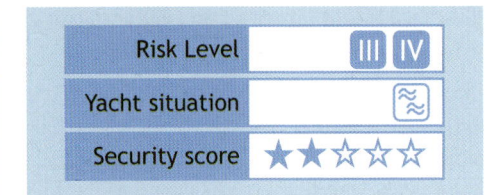

Risk Level	III IV
Yacht situation	≈
Security score	★★☆☆☆

A crew could try to use laser pointers (energy of about 2mW) to deter offensively approaching vessels at high sea during low-light conditions. These might work as they may be misinterpreted as tactical lasers mounted on a gun. Attaching them to a firearm, replica or item of similar shape might enhance the effect, at least in low-visibility conditions. During the night, they can serve as a dazzling device of limited power that disrupts attackers' eyesight (green lasers work especially well in this respect). However, although such strategies have been employed in non-naval situations and during times of civil unrest, the lasting success is too uncertain to be specifically recommended for a conflict at sea or in the anchorage.

High-energy compact lasers with an energy 1,000 times higher than your average laser pointer are a different story. Despite looking like large laser pointers, they are very dangerous devices when aimed at the eyes of attackers. Depending on the laser's colour, they are available in strength from 750mW up to 3,500mW. The beam of such lasers is very visible in low-light conditions and during the day in some weather conditions. Due to their strength and wavelength, the high-energy models' green 1,000mW lasers cause a distraction in distances up to 25NM, a flash blindness hazard up

FIG 3.34: *High-powered compact laser. (Photo: Berswordt)*

to 0.5NM and permanent eye damage up to 250 metres (yards). Looking into reflections can still cause severe damage to the eye, thus making the laser dangerous for the user if they are not protected by goggles. They can burn materials and skin at distances of up to 30cm (1ft). Obviously, these devices are not to be used for signalling under any circumstances.

There are reports of skippers passing through Risk Level IV waters who did not want to carry guns on board using these lasers to cover distances of 1.8–91m (6–300ft), with some deterrence and defence success. At present, no reports on their actual use and usefulness in a conflict are available.

High-energy lasers pose a serious threat to health and aviation if used improperly. Accordingly, some countries have imposed regulations to limit their use to trained individuals. In the USA, laser pointers must not have energy levels higher than 5mW. Consequently, compact high-energy models might not be available there. Treat them as weapons and only use them in cases of justified self-defence.

Pepper spray and mace

Risk Level	II	III	IV
Yacht situation	🥫	⚓	〰
Security score	★★★★★

Pepper spray and 'mace' (or tear gas) are very effective defensive weapons. When sprayed into the assailant's face, both products usually stop them in their tracks and put an end to all aggressive behaviour. Their method of action is severe irritation of eyes and respiratory tract, which induces a reflexive urge to close one's eyes, strong tear flow, convulsive coughing and pain. The person hit will intently focus on getting rid of the substance in the next ten minutes. Compared to many other weapons, they do not inflict permanent damage in healthy persons. However, there is a slight risk of severe injuries if someone with asthma or a severe lung disease is hit.

There have been some reports among the law enforcement community that mace is not always effective against drugged or extremely aggressive opponents. For this reason, pepper spray may be the preferred item on any boat.

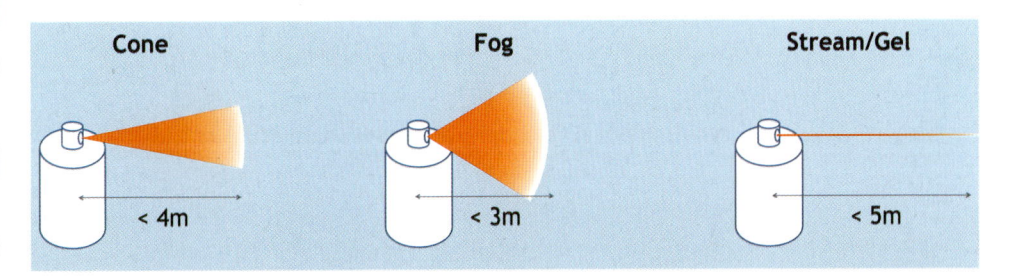

| Cone | Fog | Stream/Gel |
| < 4m | < 3m | < 5m |

FIG 3.35: *Stream shapes of most marketed products – cone and fog types are unfit for an outdoor marine environment.*

Such sprays are legal in most countries and at present no country is known to ask for a declaration when entering. It is also likely that if you use pepper spray in a self-defence situation you'll face far less hassle from the authorities than you would if you were to use a firearm. The only two drawbacks of such sprays are their limited range and their sensitivity to wind.

Despite the declarations on the containers, do not expect any spray to have a range beyond 3–4m (10–13ft). Pepper sprays come in different types according to canister size, strength and stream shape. Concerning the first two, it is recommended that you use the largest possible container and strongest formulation on board. The canister should hold a minimum of 100ml (3.4fl oz) and up to a size that you feel comfortable handling with one hand. This gives you the chance to have more than one go at hitting an assailant before your 'ammo' runs out.

In a naval and thus potentially windy environment, it may be best to choose either a stream or a gel type. These expel the active compound in a thin jet that is easier to aim in windy situations. It also reduces the risk of injuring bystanders or yourself during the action. Given comparable container sizes, stream and gel products will have the greatest range.

Especially when combined with another weapon such as a machete, pepper spray is very powerful even against more than one attacker. For this reason, it is recommended in all regions if you plan to anchor alone.

There are some discussions concerning the use of so-called 'wasp spray' against criminal attackers, but this should not be attempted because wasp spray does not have any instant effect on human beings.

Machetes

Risk Level	III IV
Yacht situation	⚓ ≈
Security score	★★★★☆

Machetes are cheap and simple but also intimidating and effective weapons. They are legal in most countries that you can visit by boat (although you need to keep it on board in some of them as you are not allowed to wield them in public). Currently, they do not need to be declared when entering a country. In the recent past, they were the most commonly reported weapons used during armed assaults on sailors.

Machetes are a very common tool in the tropics and can be bought in almost any well-stocked store, throughout the Caribbean, the Americas and Pacific countries. Compared with knives they look very impressive even in the hands of the not-so-well-trained skipper. Consequently, they are the preferred defensive weapon on some boats.

Most machetes offered in stores are gardening tools and will feature a broad, heavy blade with a blunt tip. This design makes them good at what they were invented for, namely chopping and hacking plants, but not very effective when used for stabbing, which is something of a

FIG 3.36: Latin-American bolo-style machete. (Photo: Berswordt)

limitation in the restricted confines of a boat. When looking at the victims of machete-wielding criminals, you often find serious slashing wounds that needed to be treated by professionals. Luckily, these did not prove lethal in the reported events.

If you prepare for your trip while at home, you could browse the internet for cutlass-like machetes that combine the machete's heavy blade and low price with the pointy end of a cutlass as well as saber-style hand protection. Consequently, this type of machete is quite versatile: it is more useful in defending against machete strikes and allows for cutting, thrusting and punching with the hand guard. It also looks more intimidating.

The machete should not be too long, though, since the limited space of most cruising yachts (both below decks and in the cockpit) will severely restrict their use. Medium-length ones are best.

Given that most machetes are made for use on land, the steel will not resist the marine environment very well. Prepare to maintain yours often with oil and a grindstone. Always rinse it with fresh water after beach use (to chop that firewood for the full-moon party) and, if you are on a longer cruise, stock one or two replacements.

FIG 3.37: Cutlass-style machete. (Photo: Cold Steel)

Knives

Knives – even the quickly grabbed kitchen knife – have served some skippers well during a confrontation with uninvited quests. They can be used for both chopping and stabbing motions and are a formidable and lethal weapon in the hands of a determined user, although it takes practice to be effective with them in combat.

Compared to machetes, however, they are less intimidating unless wielded by someone who truly looks like they are willing and able to use them effectively if required to do so.

They come in many variations and sizes, including kitchen tools, skipper equipment

FIG 3.38: *Bowie-style knife for camping and self-defence. (Photo: Berswordt)*

and outdoor and hunting blades as well as bayonets and combat knives. The latter are sometimes so large that they look more like machetes.

If you shop for a navalised outdoor knife, which could come in handy for camping on beaches, look for one with a stainless-steel blade to avoid quick disintegration. If you prefer to spend a lot of money on only the best quality, look for blades made from Cornidur 30 or C15TN. However, a 'normal' C440 will also be very resistant to corrosion, although it is not as capable of remaining as sharp as more expensive materials. When buying a knife especially with self-defence in mind, you might think about a tool on the larger side, not so much because it would be better in an actual confrontation but rather because it will have a greater intimidating effect on a possible boarder.

Headlamps

Headlamps have been placed in this section because they are not well suited for detection and deterrence due to their relatively low power; their light is usually ten times weaker than the beam cast by the strongest hand-held types. However, due to their hands-free design they are very valuable when reacting to any trouble on board or in the close vicinity of your vessel, enabling you to hold on to your boat or carry other useful items, such as weapons or communications gear.

When selecting a headlamp, check for parameters such as water resistance, brightness, beam distance, an alternative red-coloured light, battery life and weight.

Water resistance (minimal rating would be IP 4/4) is a must-have feature on a boat, especially if you want to use your light in normal navigation during bad weather.

The brightness of headlamps varies greatly and often relates to battery life. It is great to have a truly bright headlamp in the moment of a nocturnal confrontation since it helps to disorient an attacker. On the other hand a bright light might give away your whereabouts while you are below decks and simultaneously ruin your night vision.

Consequently, the preferred headlamp can be switched to lower brightness settings and to an alternative red light, which is the best option while you are roaming in your cabin and saloon. Make sure that the light has a memory function, which will ensure that when switched on the lamp starts in the setting used when it was switched off the last time.

Some cheaper versions always start in the brightest mode, which is not very helpful when you are getting ready for an imminent confrontation with a boarder.

A long battery life is helpful as it avoids those awkward situations when you want to check out a noise and your lamp is out of power. However, if you always have a fully charged lamp next to you, battery life and weight will not be your major security concern. These become more important if you decide to use the lamp in regular situations, such as night watches.

Higher-quality headlamps are offered in various price ranges. Given the marine environment and the damage this will do to any lamp, it might be wise to buy three cheaper models instead of one expensive one.

BEHAVIOUR AND CAPABILITIES: PREPARING THE CREW

For most skippers and crew, the last thing they want to think about before going for a sail is security; instead, cruisers prefer to focus on exploration, great sport and marvellous sunsets among the wonderful people of the world.

The good news is that as long as you are planning to sail to Risk Level I or II areas, you will not need any more preparation than common sense. When exploring Risk Level III regions, where a nightly boarding is not out of the question, less than half a day for preparation will boost your ability to cope with an intruder, thus making you more confident and relaxed.

Some boarders are robbers, whose primary goal is to control the crew. The more confident and capable you appear, the better. While the skipper and crew do not need to have the looks of a wrestling champion, a crew that does not seem fit enough to nimbly navigate their own vessel might be more likely attract criminals. If you feel physically uncomfortable running about your own vessel without constantly holding on to something, it might be wise to add crew or sail with other boats to form a support group. Alternatively, you could balance your security scales by changing your route, shunning Risk Level III anchorages or limiting yourself to marinas and monitored areas.

The same holds true for your psychological ability to actually confront a boarder without faltering. Your reactions to someone threatening or even hitting you are important, as a confrontation is as much a conflict of will as of physical force (especially as we hope to avoid any physical contact in the first place). If you feel that you will not be able to run yelling at the top of your lungs towards an equally strong boarder, you might want to practise. Most people actually have the ability when threatened, while many others can acquire it. It may seem a little silly, but it helps to train for such a situation by using simple role-playing games. Try to attack a friend and scream at him. Also practise how to stand your ground while someone runs at you: shout at him to get lost and look like you're ready for confrontation rather than flight. Develop anger instead of fear.

When a passage through a Risk Level IV area is inevitable, it is recommended that you take more time to prepare. You will find the relevant preparations for this specific situation in the section on 'Sailing in pirate-plagued waters' (see pages 130–63).

This book is not about combat tactics and it is not intended as a resource for crews who plan to deliberately linger in Risk Level IV areas. Such crews are well advised to either hire professional security teams or, if they are not from a military background themselves, get the best training available on the market.

That said, let us have a look at the first two levels of preparation that are relevant for cruising yacht crews.

Risk Levels I and II: common sense

Common sense is a good basis for a relaxed cruise in most regions of the world. Put simply, if you behave as you would in any moderate-sized Western city, you will probably have no problems in all Risk Level I and II areas, as well as many Level III areas that are not known crime hot spots.

Surprisingly, when researching successful attacks on yachts, in many cases such common sense was not used. As soon as the crews were surrounded by white beaches and palm trees their guards went down as if they were visiting a gated community in the Hamptons. Dinghies were lazily left in the water, secured by a simple bowline. The deck was cluttered with valuables, from branded surf shorts on the lifelines to diving gear and unsecured fuel cans elsewhere. Hatches and companionway were left wide open.

Close your eyes for a moment and place yourself in the urban quarter of a Western city: Rochester, Newcastle, Düsseldorf, Eindhoven, Perugia or the like. Night falls, everyone goes to sleep. The family bikes are leaning unlocked at the entrance. The garage door is open, keys in the car. The latest branded fashion items are left to dry on the clothes line in the garden. The front door and windows are wide open to allow for some ventilation, your kids' gentle breathing can be heard on the sidewalk bordering your home. Feels quite wrong, doesn't it?

Nevertheless, the same set-up is found on almost every other boat in anchorages of the Caribbean, the Mediterranean or the Pacific, places featuring populations who often face much greater material challenges than your neighbours do in the city imagined above.

Consequently, the first item on the list of a prepared crew is very simple: do not abandon your common sense when going for a cruise. Unless you are in an area well known for its safety, behave as you would in a regular city at home. Develop a simple routine when going to sleep. Pull the dinghy out of the water, clear the deck and lock the main and top hatches if they are large enough to allow someone to enter. Try to make this course of action a rule, especially when everyone is returning heavy headed from a sundowner or a potluck on the beach. The deeper you sleep, the more important it is that your boat is secured.

FIG 3.39: *Even in a Risk Level I anchorage, dinghies left in the water during the night are at risk. (Photo: Getty)*

Risk Level III areas: less than half a day's preparation for weeks of good sleep

Many Risk Level III areas are great places to visit despite criminal activities being more common there than at home. Unless you travel to crime hot spots (especially those with a recent history of gun-related attacks), the chances are low that you will be targeted by villains determined to seriously injure your crew.

Nevertheless, these regions are categorised as Risk Level III because there have been incidents reported there each year with a certain regularity. Consequently, it makes sense to prepare yourself just in case trouble does happen, so that you can relax and fully enjoy the many nights spent without incidents.

If basic fitness is a given, two hours of – actually fun – preparation will put you in a much better position to react to boarders and avoid a confrontation in the first place.

Before you get on board, it is a good idea to spend half an hour checking over your equipment. If you intend to carry pepper spray on your yacht (which is highly recommended), you should buy at least one extra canister for training. Use this in a safe environment to spray away at some simple targets, such as a piece of paper pinned to a tree. Check the actual distance that your product is able to shoot. If there is a little wind, spray with the wind from the side to see how the stream is affected.

Take some time to become familiar with the gear that you plan to use if there is an attack. Have a go at swinging your machete, put on and switch on your headlamp or flashlight, try out your communications equipment.

Start your 'naval preparation' by looking at your vessel through a boarder's eyes. While anchored, take your dinghy around your boat and identify the areas where you would try to board. Provided that you are of a similar height to most locals, see how far you can reach up on deck. Which areas are safe from a quick grab? Which sections are exposed? Try to board your boat yourself, via the sides and the transom. Where would you hold on? Where would you put your feet? How much noise do you make?

Now put yourself in the position of a potential neighbour and move to a distance at which another boat could anchor and swing safely. Which parts of your boat can they see? Where on board do you have to be to signal them? Would you be seen moving in the saloon?

For the next part, you need a friend or crew member to help you find out what a boarding actually sounds and feels like while you are resting in your bunk (or wherever you usually spend the night). This knowledge sharpens your senses for the real thing. However, just as importantly, it ensures that you don't feel worried when you're lying in your bunk and you hear harmless strange noises on your deck, such as a few birds; you will know how true danger sounds and feels and not fret too much about other noises.

Ask your partner to sneak as silently as possible from the most likely boarding points towards the main hatch or sliding doors and along the length of the boat. Listen carefully to find out whether you would hear the noises. Due to the light build of many modern boats, it is almost impossible to sneak about unnoticed. Now, repeat the procedure with a normal walk. Finally, ask your partner to simulate the sound of low-voiced talk and whispers. Ask your freind to steer his or her boat to the most likely boarding points and actually listen to see if you can hear the boat against your hull and the sound of someone stepping over. Try to notice the feeling of your yacht during the process: you might be surprised how well the weight of a grown person stepping on a boat as large as a 46-footer can be felt below decks when there is not much swell in the anchorage.

The next step aims to identify the best place for any defensive items to be situated and determine clear pathways through your vessel. In seedy anchorages, most people will sleep better if they have any defensive weapons right at hand. This ensures that they are not left completely vulnerable should a boarder manage to get inside your vessel and thus between you and your stash of pepper spray.

Test where you can comfortably store items that will help you to keep the initiative in a confrontation. Small items such as the aforementioned can of pepper spray or a strong flashlight can be placed in the alcoves next to your bunk. Larger items such as machetes might not fit close to you without posing some risk that you'll cut yourself during the night. Even if they can be stored, they might be a hindrance while you try to navigate potentially narrow passages below deck.

Run some trials of rising from your bunk, grabbing your items of choice and moving towards and through the most likely exit to confront a potential boarder. Find out where you get stuck and where you have to twist and bend to get through. Identify the best way for a fast and silent advance. On our 40ft catamaran, we were happy to have pepper spray and flashlights next to our heads and a large machete on a surface next to our companionway, since we were sure that we would always reach it on our way out before any intruder could get hold of it.

Having identified the most direct pathway to the exit, you should think about other critical positions for deterrence and defence. First, read the sections in Part IV to find out about proven sequences for defence, then define how you would react to a boarder yourself and draft your personal plan. Will you be able to reach the switch for the deck lights, strobes and foghorns without being seen from outside? Will you find them in the dark?

One very effective deterrent is the deployment of red parachute flares out of a hatch. If you consider such tactics, check whether your flares can be shot from inside in an emergency (remember, some models expel hot gas and parts from the lower end). Place at least one rocket close to you.

If you are sailing in a convoy, discuss the roles that everybody has in an emergency situation and play them through at least once.

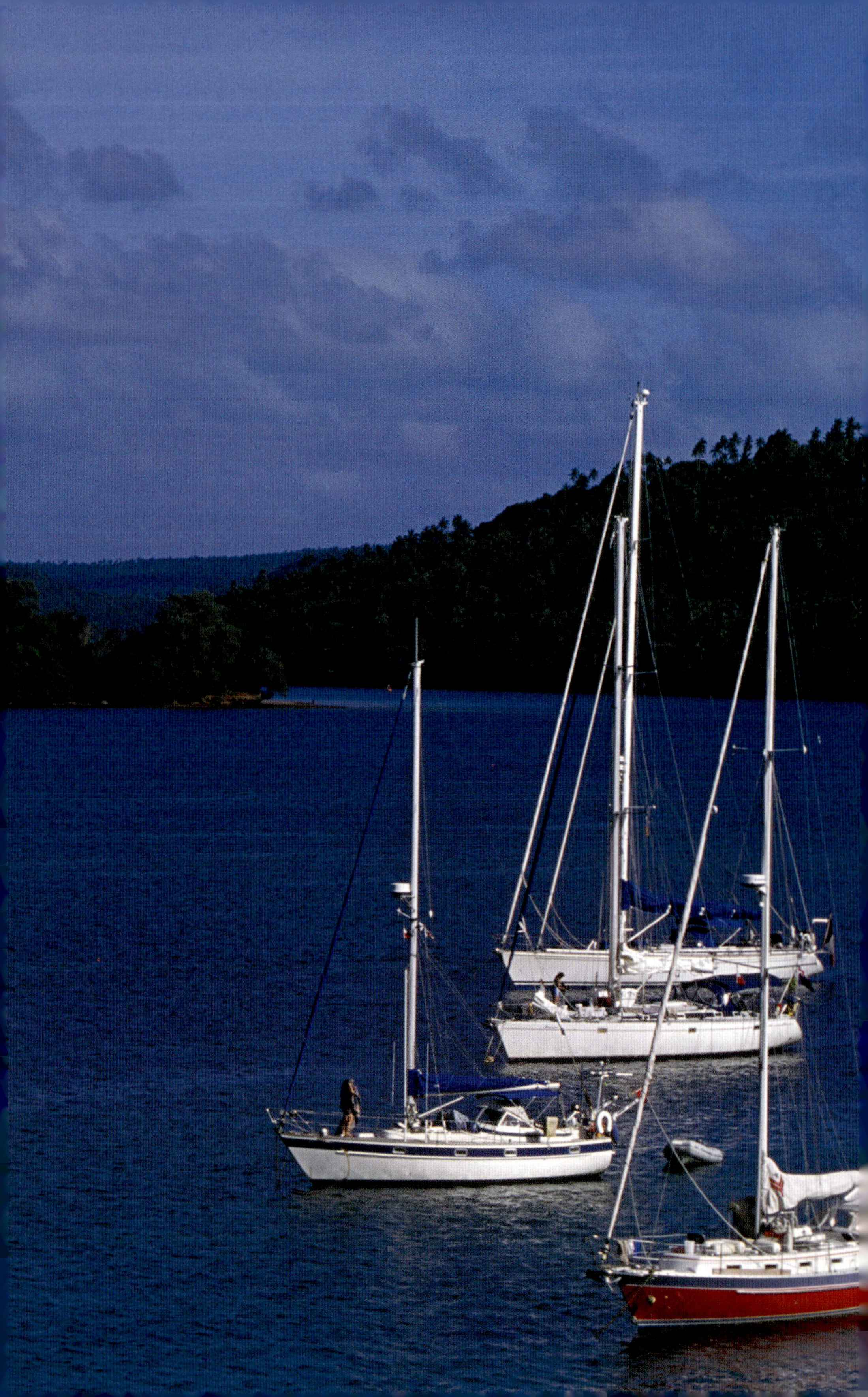

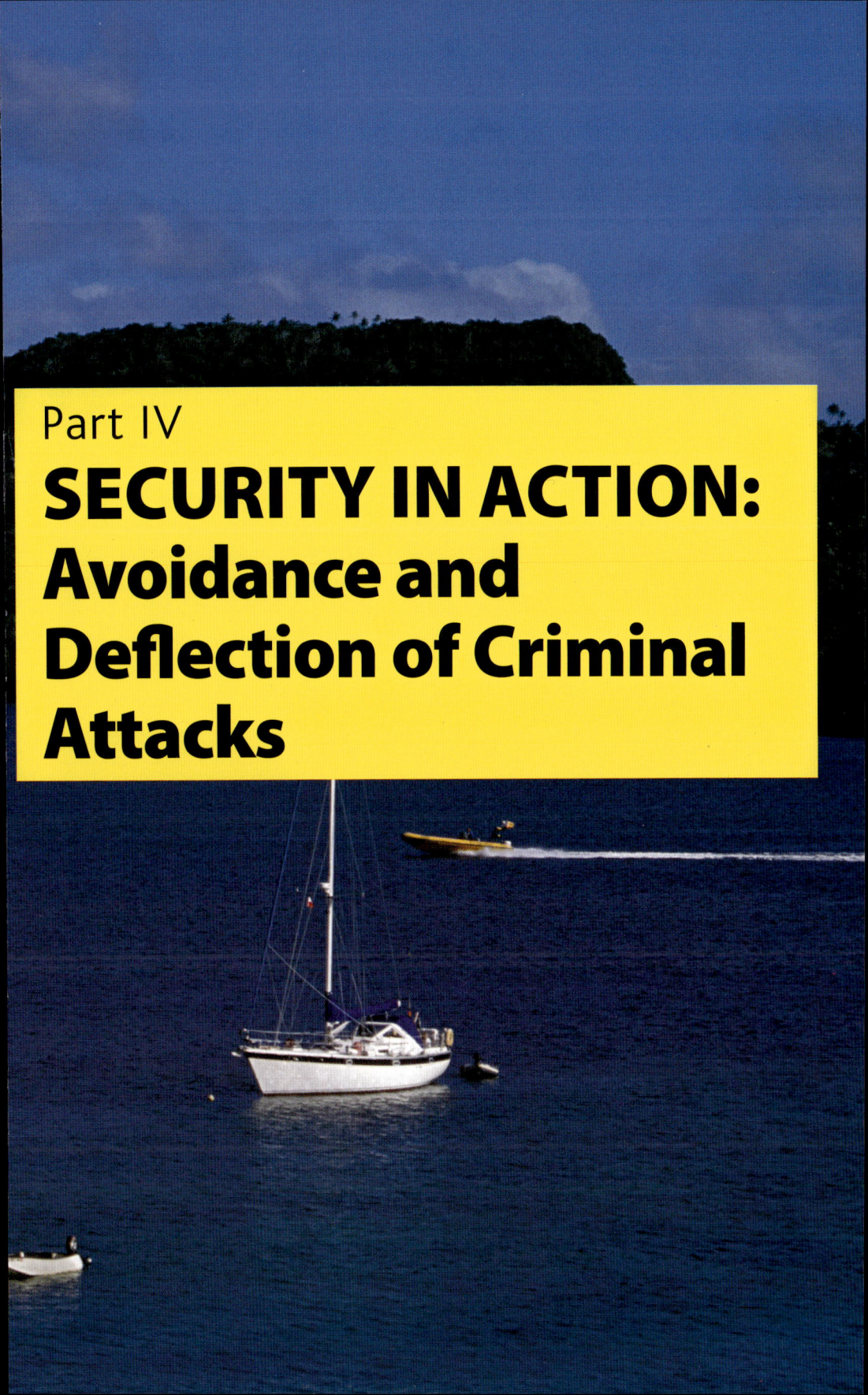

Part IV
SECURITY IN ACTION: Avoidance and Deflection of Criminal Attacks

SAFETY IN THE MARINA

> For visiting yachts, most marinas around the world are safe havens that shield their guests from the more severe crimes. Exceptions are few and far apart. Nevertheless, theft and burglary is present in most marinas throughout all cruising regions and at home. A quick bit of research and common sense will protect skippers and crews from losing equipment to local thugs.

Tying your boat to a marina pontoon, you have done a lot to greatly improve your security. While marinas may be expensive, aside from extremely rare and spectacular attacks by terrorists or daring robbers, they are safe havens even in more challenging regions: they combine the benefits of a gated community with perimeter control, access control, on-site guards and many neighbours who can support each other. Operators of marinas usually adapt their security level to the risk level of their environment. This means that even a crew who is not able or willing to risk confrontations in an anchorage can still safely explore most Risk Level III areas when they decide to spend nights in marinas.

As a result of the security measures, the threats in marinas are mostly reduced to petty theft and burglary. Unfortunately, these two remain quite common in all marinas other than the most hermetically protected ones, so you will have to prepare for theft from dock and deck as well as burglaries.

FIG 4.1: *Typical marina layout – access to pontoons secured by gates but seldom guarded.*

The following sections focus on shorter-term visits not exceeding more than a week or two. You might want to take a closer look at the quality of the fences, the guards' zeal to protect the marina and the like if you want to store your boat for a season or longer.

Before you arrive

Quick research prior to deciding upon a marina will usually give you a good impression of its security. Sources such as www.noonsite.com, printed and online cruising guides and online forums or Facebook groups covering the region are good sources for safety-related data. Marinas with a persistent security issue quickly came to light in the past.

It is also a good idea to specifically enquire about a marina's security and the type and frequency of crimes against their guests when you first contact their offices. The 'official reaction' can hold some useful information. Well, cynics would say that the answer will at least tell you whether their marketing department thought about the topic… For this reason, you should repeat the question when checking in at the front office. Usually you will obtain information about the latest problems in the vicinity as well as some tips on how to protect your valuables.

Unfortunately, almost no marina will accept any liabilities concerning criminal activities against their guests. Consequently, it will be up to you to prevent and up to your insurance to pay for any damages.

Marina design and security

More modern marinas are often built as part of a town-development scheme and are very open to the public in order to attract local consumers to the frequently included shopping malls and food courts. While this makes the marina a more interesting place for most visitors, it also invites the problems of the surrounding town, given that there is no gate security during opening hours.

In such designs, security is transferred from a single point of entry (the main gate) to each pontoon. Some layouts feature a single access to all of them, while some have an entrance to each one.

The other extreme is marinas that are either part of a gated community or protected by the terrain. Take the ONE°15 Marina in Singapore, for example. This is part of Sentosa Island, which features an access control in itself, and is located in the Sentosa Cove Area, which is gated and guarded. Within the development area, only members of the marina club are actually allowed to use facilities and restaurants close to the pontoons (no gates here, just signs). To access the boats, you need a card and to pass a vigilant guard. Unless your fellow skippers on the neighbouring yachts or the marina's staff are stealing equipment from you, there is almost no way to fall victim to criminals.

FIG 4.2: *Shelter Bay Marina, Panama, is surrounded by jungle and situated within a naval base.*

Another example in the notoriously dangerous Colon area in Panama is Shelter Bay Marina, which is situated on the Atlantic entrance of the canal. Built upon a peninsula formerly known as 'Fort Sherman' and now shared with a Panama Marine garrison, it features a main gate and armed guards all around. Although there have been some reports of theft from the boat yard, this is a very safe place in a country with some considerable security issues (as of 2017) – which just goes to show that situating a marina in terrain with an easily controlled bottleneck and with the marines as next-door neighbours does the trick.

When you arrive: get a feel for the place

As mentioned, it is always a good idea to ask the person at the front desk for news of the latest security situation. Answers could include everything from broken gates at specific pontoons to facilities or recommendations about where to park your scooter.

It is also a good opportunity to gain an impression of how they use CCTV surveillance: in many marinas, these feed pictures to a screen situated either above the front desk (for the visitors' entertainment) or behind it. Hardly anybody ever looks at those thumbnail-sized streams, night or day, so unfortunately they will most likely not be able to prevent a crime or actually raise a nocturnal guard to intercept a burglary on an empty boat. However, if they record footage, they could play in a role in catching a thief after the act.

When you have a minute, take a look at how the access points to the pontoons are secured. Compared to access control that relies on faded 'access prohibited' signs, automatic doors with locks or access cards provide added security. However, unless a guard also watches these gates they are essentially also worthless.

In many cases, especially in less-developed countries, the gates simply do not close, because either the mechanism to push them back or the locks are broken. Some gate systems feature magnetic locks: a brisk shove will open these without a card. However, it is we – as the users – who cause the worst security problem. You will understand this once you have watched such doors for a couple of minutes.

Due to the general friendliness of yacht crews, they will literally hold the door open for any person – authorised or not – while getting off or on to a pontoon. The same holds true for mechanics and service crews. Everybody will hold the door for following people, something that has saved my day many times when I have forgotten to take my key card while venturing outside, but which isn't great for security.

Consequently, whenever you reside in a marina without a main gate and guards on the passage to the pontoons you can assume there is no access control whatsoever. There will probably be a lot of theft and possibly burglary, so you might as well prepare as if you were moored to a Mediterranean town dock.

Avoiding thieves and burglars

Marinas work much like smaller communities: some crooks may threaten the security from outside, but there may also be some problems with external or internal staff as well as fellow boaters. With this in mind, you need to follow some simple rules to prevent any issues.

Never leave items on the dock itself when you are not around or while you are sleeping on board. If you use bikes, scooters, longboards or the like for your transportation, you should lock them to a fixed point on the pontoon or take them on board. These are the prime targets in a marina. Make sure that no unsecured items can be grabbed from the pontoon or from a boat in the water; it is best to keep your deck clear when you are not around.

If your outboard is accessible from the water, make sure you secure it with a high-quality lock and chain. Although only the boldest thieves would carry an outboard through the gates, there have been reports of criminals arriving and leaving by boat.

Several skippers have installed warning signs on their main hatch announcing alarm systems to deter burglars. Research indicates that this resulted in no attacks being carried out on such boats, although the sample size was very small.

Reliable alarm systems can be very helpful against burglars. In contrast to a CCTV surveillance system, a siren howling away will most likely summon guards and neighbours. Before arming a system when leaving for a longer period, you must ensure that it will not produce false alarms or you will be the most hated person on the pontoon and may face some claims for expenditure from the guards, police and technicians who had to disable it while you were gone.

MOORING BUOYS AND SOLITARY BAYS

Most risks are eliminated by choosing a known safe anchorage in the first place. However, exploring uncharted anchorages is no issue at all if you follow a simple sequence from approaching the coast to preparing for the night. A couple of preparations and entertaining exchange with the coastal community will help to prevent crimes against your yacht.

As long as you are not stopping in areas notorious for gun-wielding criminals, uninvited visitors will most likely carry clubs, knives, machetes or – in case of thieves and burglars, no weapons at all.

In the very rare event of a nightly boarding, many skippers have successfully thwarted an attack carried out by criminals who were not carrying firearms. Tactics that have been employed successfully in the past include quickly assessing the situation and then seizing the initiative to swiftly and aggressively confront the criminals until they departed over the sides.

Warm nights under a fabulous starry sky in a bay fringed by lush forest, with the silence only disturbed by the occasional call of a mysterious nocturnal jungle dweller: for most sailors, this picture is the prime motivator for embarking on a cruise.

However, as discussed in the 'Planning Your Cruise in Uncertain Regions' section (see pages 36–47), most attacks on yachts are made against boats at anchor or on mooring balls. This makes a lot of sense from a criminal's perspective, given that anchored boats are more easily found and approached than vessels cruising offshore. At the same time, anchorages and many mooring fields lack the tight security and support network of marinas.

Consequently, sailors need and desire to find a way to visit and enjoy such dream destinations without converting their vessel into a fortress or spending their sundowners and their nights with a constant sense of uneasiness.

Unfortunately, most mooring fields – even if they are associated with nearby marinas – usually do not offer more protection than an anchorage. Unless the operator of a given field explicitly offers protection and nightly guards, it is wise to prepare for the night in a mooring field just the way you would in an anchorage.

The most important strategy to ensure relaxed nights at anchor (or mooring, for that matter) is to go to well-researched places. If you are certain that the expected risk matches the skipper and crew's self-esteem and preparation, everyone will feel good and less stressed. Remember the security balance from Part II and avoid any stops where the risks might outweigh the security measures or your vessel and crew.

Let us take a look as some tactics that will help you enjoy a stopover in most bays, from

Risk Levels I to III. Before starting, once again I warn you: do not consider anchoring in Risk Level IV or III areas with a recent history of attacks utilising firearms.

Approaching the anchorage

Ideally, the skipper will already have an idea about the security situation in the anchorage before he approaches it. Prior research will help to predict or interpret anything that happens at this stage.

Take a simple example: a boat approaches a secluded bay with a tricky reef passage at dusk. All of a sudden, five local boats approach in the settling twilight, some rowed, some with rattling outboard engines. The approaching mariners shout, signal and come dangerously close, while the crew tries to negotiate the tight entrance to the anchorage. There is really no time to carefully listen and understand the heavily accented language. The situation is chaotic and disturbing for uninformed skippers. Perhaps these are boat boys or other sea-borne service providers, possibly they are concerned coastal dwellers who want to protect the fragile reef from anchors. Maybe they are aggressive beggars or young, drunken guys having their kind of fun with a rich visitor on a yacht.

In many instances, the general atmosphere as well as the behaviour of the coastal community can be quickly researched by checking with the locals of the current anchorage, the internet, and other sailors who know the place or have visited before. A good coastal guide can also help, as long as they are not outdated.

Depending on their knowledge and ability to put the visitors behavior into perspective, skippers should either feel much more relaxed in this situation or avoid the place altogether.

Whenever you intend to stop at an entirely unknown anchorage, it is very wise to plan the approach so that it occurs at a time that leaves open the option to sail to an alternative place within daylight hours. It is great to not have to stay in a seedy place simply because it is too late to move somewhere else.

Before leaving deep offshore waters to close in on an unknown coastline, skippers should make sure that the track feature on their GPS equipment is activated. This will come in handy should the crew decide to leave an anchorage at high speed during the night. If tides are relevant, it is a good idea to take a route that would be passable even during low tide so that in an emergency you know you can always make a get-away.

If there is no recent and reliable information, any anchorage should be approached at a slow pace, for several reasons. First, by approaching slowly skippers have the time to check the beaches and hinterland with their binoculars. Is there a village? Who is at the beach and what are they doing? Activities such as bathing, playing, fishing or working are good signs of a healthy and usually safe community. The more women and children are present, the better (provided that the visited society permits the public appearance of women). Groups of men are no problem in many regions if they seem to be engaged in productive activities that are common for the region. In the Pacific, these could be fishing, collecting clams,

	Positive	Suspicious
On The Coast	✓ Women and children present ✓ Males engaged in productive activities ✓ Well-maintained dwellings, structures and gear ✓ Well-maintained beach	! Groups of loitering males ! Signs of alcohol or drug use without an actual festival going on ! Shabby buildings and structures, general signs of neglect ! Garbage-strewn beach ! Signs of contraband crops in the hinterland
Behaviour	✓ Hailing gestures, welcome calls ✓ Approach to welcome and chat ✓ Approach for bartering or trade ✓ Seamanlike conduct towards visiting vessel and crew ✓ Reasonable reaction to friendly requests to keep a distance while manoeuvring	! Waving off and aggression ! Aggressive approach intended to intimidate ! No reaction to friendly request to keep distance during manoeuvres ! Requesting money or fee without explanation ! Bumping into vessel ! Fixing lines to vessel without asking for or receiving consent

FIG 4.3: *Getting a feel for an unknown anchorage: positive and suspicious indicators.*

repairing nets, etc. Groups of loitering men, possibly drinking or using other perceivable drugs might make the crew suspicious and raise the level of caution.

Going slowly also allows the coastal community to react to the newcomer and gives the crew sufficient time to react in return. In many bays in remote areas, a yacht will be hailed by waving and friendly welcoming shouts. Kids will jump into the water to swim toward the visitor and perhaps some young men will approach with their boats for an early chat or barter deal. Skippers now have the time to get a feel for the place and the crew can look for signs of security or possible threats. Are the people friendly and respectful? Are they curious about new people? Do they want to do business (bartering or trading)? Do the buildings look well maintained (even the simplest hut can look great or shaggy)? Is the beach clean?

Most fellow mariners will respect your course and keep a safe distance away until you have dropped and secured the anchor. All of these are good signs, even if the approaching boaters are a little persuasive.

Signs of higher risks are boaters who do not respect the yacht's course and threaten to intercept or even bump into it. This is especially the case if the mariners are not reacting to friendly but firm requests to stay away and there does not seem to be a corrective element (such as others in the group opposing the culprits and helping the visitors). In this instance, the crew should consider their security measures or simply leave.

In some regions such as Melanesia (e.g. Vanuatu, Solomon Islands, Papua New Guinea), local culture defines the ownership of a bay and reef as property of the tribe or community living at its coast. Consequently, it is an acceptable custom to be asked for a contribution, which is usually given to the chief of the community; in most regions, welcome committees do not collect such fees. If someone asks for anchoring fees at this early stage, it helps to announce that you would like to discuss the matter with the chief. If this is denied, ask for a receipt (often impossible to produce) or to be allowed to take a picture of the cashier, which usually resolves the matter; otherwise, the crew is about to be cheated.

Generally, it is always a good idea to simply follow your instincts. If the place feels seedy, it is better to leave (or prepare to install beefed-up security for the night).

Choosing a spot to drop the anchor

Most of the time, the prime criteria regarding where to drop the anchor in any bay chosen for the night should be seamanship and comfort. Given that some places in an anchorage are equally suitable in this regard, security considerations should now also come into play. These might become more important if a crew decides to stop for the night despite the presence of adverse factors. The first ones listed below are especially important if the vessel is alone in a bay.

One security aspect neatly harmonises with nautical and comfort requirements: the less sound there is from wind and waves, the better the spot. Fewer noises around the boat and in the rigging leads to sleepers who will be woken by any suspicious noises more easily. These include approaching boats, nearby voices, skiffs banging against the sides or footsteps on deck. This is really important – several victims of night-time robberies have reported in the past that they didn't hear their attackers until they entered the cabin due to the noises outside. On the other hand, many woke to nocturnal engines and voices 'next to the bunks' in the night and were consequently able to confront and chase away the would-be boarders before they gained the initiative over the crew.

Another critical security factor is distance to the beach. Generally, a yacht is safer the further away it is from the shore. When analysing crimes against anchored yachts with crew on board,

FIG 4.4: *Gizo, Solomon Islands, population approximately 5,000 – just a few too many people for it to be trouble free. (Photo: Berswordt)*

it becomes apparent that 30 per cent of the attackers approached swimming. The greater the distance, the fewer swimmers will be motivated or capable of reaching the vessel.

This simple distance principle may be overruled in cases where the crew is able to engage in friendly contact in their bay or where they are invited to join the security of a group or organisation close to shore. The spot next to the yacht club in the bay of Rabaul, Papua New Guinea would be one example of this scenario. Invitations from village heads in the Vanuatu, West Papua or the Solomon Islands are others.

If there is a larger settlement – anything larger than a village – at the shore, it is usually wise to keep a distance from it if the anchorage allows this. Even smaller cities with populations larger than 3,000 attract and produce social problems in less-developed parts of the world, including unemployment, a lack of social bonds and drug abuse.

A good measurement is the ability to engage in personal contact with inhabitants. If the settlement is so large that a sense of anonymity prevails, it might be the source of problems and less of a source of support if a situation were to develop.

If other yachts are present, it is always a good idea to flock together, if seamanship allows. When choosing to anchor with others, skippers should keep in mind that any distance greater than about 45m (150ft) will diminish the added security from anchoring in a group. The worse the weather or the louder the environment, the closer the boats should be.

Exploring and connecting: the coastal community

Apart from the most remote areas and islands, most coastal regions are populated in some way. Settlements range from multi-million complexes to single houses to temporary fishing camps.

Many long-range cruisers and circumnavigators agree that getting in contact with the population ashore is both fun and serves security purposes in areas where the specific risk level is uncertain.

Contact often starts when ships are approached while going into an anchorage by boat boys. If the situation permits, the crew could slow down or stop their yacht for a chat. Is it OK to anchor? Do they have items to sell or trade? Where do they live? Where is the village? Is there a market where you can top up provisions? Where is the best and most secure place to land with a dinghy? In some cultures of the Pacific, a visit to the settlement's chief is customary for new arrivals. Respectively, crews might announce their plans and ask where he can be met.

Besides acquiring useful information, by having these conversations the crew will gain a first impression of their potential neighbours. Are they friendly? Are they simply looking for a quick buck or are they genuinely interested? Are their offers fair or just a plain rip-off? Do they respect the boat and its crew (keep a safe distance, do not constantly bounce into the hull, etc)?

FIG 4.5: *Visiting families and working boaters are signs of an intact, secure environment. (Photo: Berswordt)*

Some inhabitants of the coast in remote areas are, judged from a monetary perspective, utterly poor. They dress in very ragged items and use wildly improvised vessels, oars and gear. Such looks are not at all an indicator of higher threat levels; rather than judging by the looks, crews must focus on behaviour. Smiles vs closed faces or aggression, relaxed demeanour vs agitation and pushiness, tranquil self-esteem vs the flicker of too much of the local fad in stimulants. Keep in mind the fact that around the world there are many people considered extremely poor by Western standards who appear to be much happier and more contented than we have ever been.

When the offers of traders are fair, consider buying or trading items even if you do not need them. A successful transaction promotes you from stranger to trade partner. You have thereby contributed to the benefit of the community, which builds bonds.

Not all bays with a community of pushy, persistent naval salesmen offering trashy tourist items are dangerous. However, after conversing in a friendly tone (sometimes being more serious to keep those metal canoes off your precious gelcoat) and smilingly rejecting their offers of China-made bogus handicraft, you might consider it a good idea to vamp up security a little. In the past, we felt that this was especially indicated when the welcoming committees were obviously telling lies, perhaps concerning their clearly mass produced 'local handicraft', possibly to lure us into paid anchorages or overpriced private mooring fields, despite the available anchorages being safe, more convenient and eco-friendly. Although it is a long stretch from selling trash for an exorbitant price to stealing, the vibe is mildly alarming. If the conduct towards the visitors persists on shore, the crews should be well prepared during the nights.

One of the upsides of getting to an unknown anchorage early is that the crew has some time to visit the coast and make friends with the locals. Perhaps the crew have already

FIG 4.6: *A small coastal market in Melanesia. Get ready to mingle and hear the latest gossip. (Photo: Berswordt)*

collected recommendations to visit certain places or people from the boat boys: a sister who sells fruit, a brother who has fish on offer, or the favourite local watering hole.

When taking in the beach scene, check again for loitering groups. Don't mind some men relaxing after a tiring fishing trip. However, those seemingly doing nothing and simply killing time are indicators of a community that is not 100 per cent intact. Conduct a brief scan for overt use of drugs and alcohol in the community.

Attempt to identify trustworthy people who live right at the coast. These are often fishermen or people who have their houses or gardens at the waterfront. Introduce yourself and try to get chatting with them. A little small talk about their boats, house or garden melts the ice in many places. Ask them for advice about the anchorage, the place where you pulled your dinghy up, etc. Smile a lot. Perhaps they can sell something that you need or can afford. Just as with the boat boys, trade builds bonds and people often tend to help others with whom they do business (or simply like, for that matter).

Such contacts are a lot of fun and one of the reasons why many of us set sail in the first place. However, they also serve security purposes in two ways: on the one hand, people who get to know you are more likely to warn you or call the police when your vessel is approached by possible offenders; and they may also warn you of any security risks in their community. In most cases, these are vague caveats concerning the general situation ('watch out, there are a lot of thieves around!'), but sometimes they are precise alerts concerning specific areas or groups of people.

FIG 4.7: *Friendship is the best insurance. A tender from* Alytes *helps to bring villagers to shore from SY* Kings Legend*, which picked them up from a remote island after the village's transport vessel had critical engine failure. (Photo: Berswordt)*

A report of a spectacular case of positive 'intel work' was posted in late 2015 by the skipper of SY *Stella* on www.noonsite.com. The skipper of a local boat had invited him over for a coffee. During the chat, the local warned the Westerner about a rather new recruitment camp of Abu Sayyaf (a terrorist group notoriously known for abducting Westerners for ransom). This camp was on the shore of the very bay where they were currently enjoying their coffee – this was surprising given that they were more than 560 kilometres (350 miles) from the group's home ports. The information was most likely a lifesaver for the visitors.

Preparing for the night

Besides fixing that sundowner, some quick preparations can be made to ensure a quiet and relaxing night. Some of these will not be necessary at all in most anchorages, while others have proven helpful in insecure areas of Risk Level III regions.

The first thing to do is pull up the tender: every place, every night, no excuses. The dinghy and outboard motor are the prime targets for any thief anywhere in the world. Depending on the system used to pull up the dink, as well as the perceived threat from thieves, it might be wise to lock it to the boat with a good padlock and chain.

If not done after dropping the anchor, the crew should clear the deck before retreating below. Nothing should be left in the open. Focus in particular on things that are accessible

FIG 4.8: *An unwise set-up for the night in the port of Djibouti: fenders can be used to help boarders climb up and laundry can be snatched from a boat without the thieves even having to board the yacht. (Photo: Berswordt)*

from boats going alongside or even swimmers, as well as anything that could be used as a burglary tool or weapon. This is recommended for all risk level areas. Items that cannot be stored and locked away (such as the notorious water and diesel tanks on small cruising boats) should be firmly locked to the boat with chains and padlocks.

For a comfortable night, all sounds should be silenced so that any inexplicable or threatening noises can be clearly heard. The squeaking boom should, therefore, be tied to the side, all halyards secured away from the mast and fenders, hooks, paddles – everything should be stored away to prevent it from moving on deck or in the cockpit. On a silenced boat, the only strange noises will come from an intruder. This ensures a good night's sleep for everyone on board.

You should also close the lifelines. This is not a security measure, but rather signals that you are not welcoming visitors right now. In some regions, curious coast dwellers, especially kids, might simply come on board to say hello. A 'closed door' is usually universally understood and prevents misunderstandings.

If you are in truly challenging areas (Risk Level III), you could now rig warning signs and, before retreating to the inside, switch on external lights and camera systems that might be on board (see the section on 'Security equipment on board', pages 55–101).

Close and lock all doors and hatches. Although nightly boardings are very rare in Risk Level I and II areas, it is still recommended that you don't easy access to the interior of the boat. Those with air conditioning, removable bars at the main and deck hatches or hatches that are too small for entry are fortunate. In areas with a history of crime as well as in anchorages that the crew deems to be seedy (all risk levels), there should be no open and no unlocked access to the inside of the boat. Not doing so has resulted in many crews not only losing the initiative to resist but also being awoken at knifepoint, allowing even the most untalented robber to steal valuables.

If feeling truly uneasy, skippers might choose to leave the chart plotter on in case the crew wants to leave in a hurry, using the inbound track in reverse as a guide for a safe passage out. In areas where radio watch from police, coast guard or other authorities is expected, the VHF should be turned on. If there are other boats in the anchorage, the radio might be switched to an emergency frequency agreed by the group in advance.

In areas where nightly visits are to be expected, at least one crew member should have their favourite defence equipment ready to grab and use in a possible confrontation. A clear head is also required in these situations, so it is best to forego those sundowners if there is a serios threat of visitors at night.

Sleeping the night away

As it should be, spending the night somewhere will be a refreshing pleasure if the boat is anchored well prepared in a soundly researched bay. If the deck is clear and all sources of noises are fixed or removed, the crew should be able to relax and sleep safe in the knowledge that, having prepared a little, they will wake up if they hear suspicious noises, even if there is no guard awake. Obviously, if alarm systems have been fitted, these will help to raise even the soundest sleeper.

Only a very few places are so dangerous that they require an anchor watch due to security concerns. Usually, crews simply avoid going to these places. However, if forced to stop in unfavourable location or if the on-site situation looks worse than expected, a situation could arise where a watch might be necessary.

In high-risk areas, sleeping on deck should not be confused with keeping watch. Quite the contrary: sleeping on deck simply means that you are more easily accessible and in a more vulnerable position. There are a number of cases of deck-sleepers being surprised and swiftly controlled (or shot, for that matter) by robbers. Consequently, guards should be awake on deck.

Hand-held radios are great for communicating with the crew below. Set to a low-traffic channel and low transmission power, these do not disturb the sleeping crew when all is clear, yet, if there is an emergency, the crew can be raised quickly and emergency frequencies can then be used to call for assistance from other boats or the shore.

Confronting and fending off boarders

If you have researched your anchorage and prepared your boat, the chances of a nightly boarding are very low. Nevertheless, there is always a first time in previously safe areas and

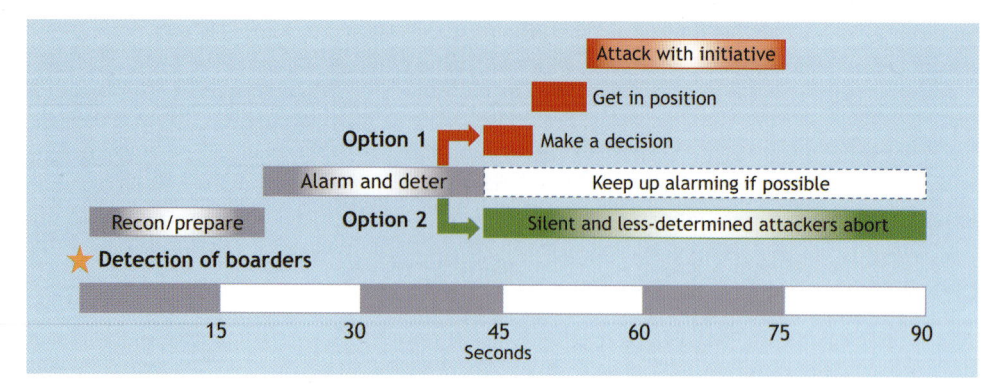

FIG 4.9: *Sequence from detection to neutralisation of a boarder. Initiative and offensive action are key.*

sometimes you could be forced to anchor in places that you would usually avoid.

This book does not cover the tactics of a firefight between two parties armed with guns; instead, we focus on how to chase off unarmed or lightly armed criminals with equipment conveniently available on a yacht.

Before proceeding into the details of a confrontation, here's a quick summary of the most important findings described in the 'Crime dynamics: a introduction to avoiding and repelling criminal attacks' section (see pages 17–28):

- If you are not stopping in a region notoriously known for gun-wielding robbers, the boarders will most likely carry melee (hand-to-hand combat) weapons such as clubs, knives or machetes. There is a high probability that they will be unarmed, because they are thieves or burglars, not robbers.
- If you are confronting the boarders with initiative and maintain offensive action, the chances are high that you will successfully chase them off your boat without being injured yourself.

This is good news for most of us in most regions of the world.

Let us first think about initiative again. If you have stuck to the recommendations made in this book, the boarders will not have entered the cabin or been able to control you from the outside before you are aware of them. As a result, the initiative is neither theirs nor yours. All the following actions you take should aim at claiming the initiative and then deciding whether to confront the boarders or to barricade yourself somewhere, call for help and try to sit it out in a safe environment. The latter strategy could be wise in instances when you realise that you are utterly outnumbered and out-gunned (or out-knived, for that matter).

It is best to detect the boarders before they get on deck. Outboard motors closing in at night or the bumping noises of a boat against your hull should always prompt you to find out what is happening. Often, would-be criminals will leave if they are detected before boarding. Pose with lights and weapons. Raise the vessel's alarm systems or make a lot of noise if suspicious people keep approaching despite you being on deck.

If boarded while below deck, the main goal of a crew is not to engage in any kind of physical conflict (including firefights) but rather to chase the offenders from the boat. Do not try to arrest a criminal. The night may be long and hosting retained captives is complicated and not without dangers: it is simply not worth it. Do not waste time taking pictures or collecting evidence. Simply concentrate on getting the boarders into the water as quickly as possible.

Confrontations are very short affairs. It is probably not overstating the case to say that the outcome – whether you will be a successful defender or beaten victim – is decided in the first 45 seconds of a boarding. If reacting properly, the whole sequence usually lasts less than two minutes from the moment the crew notices something is awry until the attackers are overboard, or the crew is controlled and ready to give up their valuables.

Time is a capricious ally during a confrontation. As long as you have not effectively called for help – ideally with the offenders noticing these calls – time is on their side. They will use it to get their bearings on your boat, occupy advantageous positions, cover escape or attack routes, pry open the main hatch or carry off items from your boat. On the flip side, from the moment the crew has raised alarms (and others are around to potentially support the yacht), time is on the defenders' side.

Nevertheless, unless you decided to barricade yourself inside the vessel, it is a good tactic to deny the assailants every possible second on your yacht. Act quickly and aggressively. Most boats do not allow for smart ambushes with the crew ready to jump at intruders from dark corners. The longer you take, the more likely it is that the criminals will cause harm.

When looking at successful defences carried out by courageous crew in the last five years, some basic tactics in a rough sequence have proven effective.

Step 1 – Recon and prepare

Try to obtain as much information as you can about the boarders in the first 15–20 seconds.

- How many are there?
- Where are they right now?
- What equipment have they got?
- Are they carrying guns?
- Are more boarders to be expected (is there a boat with further crew alongside)?
- From where will these reinforcements enter your boat?
- Does the weather permit aimed pepper-spray deployment?

Remember the types of attackers described in the introductory chapters. Thieves and burglars, especially if they are inexperienced, often work alone. They will most likely be unarmed (you should chase them off aggressively, regardless). Robbers, however, usually attack in small groups of two to four. For this reason, assessing how large the group of attackers is gives you an idea about what to expect.

Use the time to equip for a confrontation. If well prepared, you will have at least a can of pepper spray, a strong light and perhaps a machete at hand by the end of this step.

Step 2 – Alarm and deter: raise general hell

This is undertaken from covert positions and serves multiple purposes. Primarily, you signal that the attack has been noticed (a problem for silent attackers, such as thieves and burglars) and the offenders have lost the element of surprise (a problem for controlling attackers, such as robbers and kidnappers).

You also alert potential supporters, thus setting the clock running against the criminals. Depending on the equipment that you have to hand, you will turn the odds in your favour by putting light on the boarders while staying in the dark yourself. Some equipment will also startle or unsettle them.

Useful equipment during this step includes hailing devices/sirens, bright deck lights, navigation lights, strobe lights, the DSC distress on your VHF and a red parachute flare safely shot vertically from a covert position. Try not to give away your position in the vessel at this time. Depending on the environment in the anchorage, the offenders (burglars, thieves and less-determined robbers) might jump off your boat at this moment: they are illuminated, others might notice the commotion and they have no one to control at this time.

A panic button or a remote control for your fog horn is very useful in this situation as you can use them to raise the alarm from the relative safety of your cabin.

If you have more than one or two crew members on board, keep up a constant noisy and bright environment. Many systems can be configured to continuously make noise or light once activated. This is very helpful for the defenders.

Step 3 – Decision making

Take a very short time to make a decision. Will you attack the boarders or will you try to barricade and hope for help?

Remember: if professional pirates or terrorists have boarded your vessel, it is very unwise to resist at this stage, even if you are armed with guns. It is better to try to either escape or barricade and call for help.

Despite some audacious crews successfully resisting gun-wielding criminals without carrying firearms themselves, this tactic is extremely risky as your life is at stake if your plan fails.

Step 4 – Get in position

When and if you have decided to confront the boarders, you may first have to gain access to the deck. Depending on your gear and yacht layout, this might be difficult without giving away your position before you are actually on deck (think of noisily fiddling with the lock of your main hatch while trying to surprise the boarder). Depending on the assailants' position, it could be better to emerge from a hatch rather than the companionway. Approach without warning from an unexpected location, if possible.

Step 5 – Attack the attackers with initiative

When there are boarders on your boat, there is no room for niceties: no discussion, no arguing. When you choose a confrontation, play to win. Gain, use and keep the initiative. Boldly move in a way that clearly leaves only one direction in which the criminal can go: over the sides.

To achieve this, you must move quickly, loudly and aggressively. If surprise is on your side, approaching determinedly toward the attackers while shouting at them to get off has worked in many reported cases.

Do not run, but rather walk briskly and in a controlled manner. You do not want to trip over a line while charging. Command them to leave at once during your advance and throughout the entire confrontation.

Have the most menacing weapons you own with you. If this is a firearm, this is what you might consider carrying. Remember, however, that there is almost no scenario in which you will be able to use it legally in a foreign anchorage. Consequently, you need to carefully think about even firing a warning shot. Remember: in many scenarios, guns might cause more trouble that than they solve.

If confronting assailants who carry anything less powerful than a gun, the combination of pepper spray and a machete is the most convincing combination, short of a gun. Machetes look intimidating and are at least a match for most offenders. Pepper spray gives you a stand-off capability, which puts you in a superior position against any attacker packing melee weapons. A strong light will also help to blind and dazzle the attackers.

The moment that you enter into range, use pepper spray against the boarders, if the wind permits. This will effectively end most confrontations. With initiative and surprise, a single person armed with a large pepper spray container can quickly defeat a group of four of five assailants who are not carrying guns. Remember: do not hesitate, do not argue. You are clearly in a situation of self-defence.

If they are incapacitated by either your stand-off attack or a melee strike, command them to leave or get them overboard if they do not move by themselves.

If the attackers do not yield, you can and should use reasonable force as soon as you are sufficiently close, especially if they are armed. Bear in mind that a good shove in the right direction (the water) is often more effective, and less legally fraught, on a narrow-spaced boat than striking repeatedly with or without a weapon.

Your goal is to chase the boarders away, not to injure them. In most countries, the law allows for a reasonable reaction to an ongoing threat. Hacking away at the head of an unarmed burglar who is about to jump over the sides is not reasonable. Shooting a knife-wielding thief in the head from a distance is equally not considered reasonable in many countries.

Bear in mind that at the end of the confrontation, you will most likely face two people: a police officer and your image in the mirror. Acting in a determined but reasonable fashion will make both encounters a lot easier.

Step 6 – Managing the aftermath

Directly after an attack – regardless of whether or not it is successful – the crew should assess whether they want to stay or leave quickly.

When you are anchored alone and there isn't an effective support infrastructure ashore (no police or coast guard post, no reaction during the attack, no friendly faces the crew has met the evening before), leaving could be a good idea, especially if you have to fear that the attackers will return in greater numbers and better armed.

If you are anchored in a group and if the shore community reacted supportively to your struggle, it might be safe to stay. After all of that commotion, doing so will definitely be the more comfortable option, since an emergency departure will most likely lead to an unplanned night at sea when you are already stressed and upset.

Whatever you do, it is best to stay on board until daylight. Do not try to go to the police at night. Doing so will put both yourself and the boat in danger, especially if your crew isn't sufficiently large that you can leave a well-sized guard team on board. Instead, you might try to call for help with the VHF or your hailing system. Depending on the motivation of local authorities, someone might actually come out to you. The chance for nightly support will be greater if there are navy or coast guard units in range. In contrast to most local police staff, navy and coast guard units tend to listen to VHF channel 16.

Do not forget to take pictures of any damage and make a list of things that might have been stolen. This will be helpful for your insurance as well as local law enforcement.

In some circumstances, attackers in the past have left items of their own at the scene. Some of these were valuable to their owners, such as, in one reported case, a dugout canoe. It is recommended that you get rid of such items, especially if they can be seen from the shore, since they might pose a powerful motive for the criminals to return to your boat for a second raid. However, before you discard them it is a good idea to take pictures that could help the police to identify the owners.

Involving the police is generally a good idea the next morning. The reports on local authorities' reactions to such contacts vary widely – from a mere shrug of the shoulders to very effective documentation and personal support for the crew. Although there seem to be local differences, it is very difficult to predict whether the visit by the authorities will help or not.

If you find the time, it would be very helpful if you could post a report on www.noonsite.com. You will find a good format for doing so in the website's appendix. These help others to understand the risks in the area where you have been attacked, and the report will also be a great contribution to a better understanding of criminal threats to cruisers in general.

Emergency departures

An emergency departure is defined as an unplanned and quick departure from any anchorage or mooring buoy to the relative safety of the open sea. It could result in a night on the open water and is usually conducted under engine power and with all navigation lights switched off.

When deciding to depart in such a manner, skippers will be very happy if they have prepared by recording a track of their safe arrival path since this makes navigation easy: simply follow your entry track out again (taking into account tidal changes).

There are several specific situations in which an emergency departure is probably necessary. One is after a crew has successfully repelled an attack but fears retaliation or a better-armed rerun of the attack featuring greater numbers. This is probable in regions where there is no supportive network (other boats, the police, helpful locals) and/or the local community seems collectively hostile or totally unconcerned.

The other situation reported several times is developments before or instead of an actual attack that prompt a speedy exit; when strange and unsettling activities develop ashore or in boats in the anchorage, it might sometimes be better to leave than to wait for an escalation. Such events could be massive and collective abuse of alcohol or drugs with clear signs of aggression, continuous 'patrols' of unfriendly boaters around your yacht or serious warnings from trustworthy shore dwellers.

Going on shore excursions

Despite being marine explorers, we must admit that some of the greatest discoveries and encounters could await inshore.

Leaving the yacht for a longer period leads to two problems: where to leave the dinghy and what to do about the boat.

In regions that are notorious for burglary and theft, most attacks on both tenders pulled up a shore and anchored yachts are conducted during the dark hours. Consequently, you can strongly increase security by ensuring you are back before nightfall in such areas.

Unless you know that you are in a safe environment, the dinghy should be left in a guarded place. Possible positions could be a yacht club's dock; a customs, police or coast guard dock; or a beach in plain sight of a bar or restaurant. Restaurant owners are usually happy to keep an eye on both the dinghy and boat if you have a coffee and food at their establishment.

Unless you tie up to the authorities' pontoon, you should always lock your dinghy to a fixed object. Simply pulling the tender up a deserted coast is the riskiest way of leaving it. If you must do this, try to get to a fixed object such as a signpost, massive rocks or trees to which you can lock the tender.

If you stay away for more than a day, it might be better to seek alternative transportation from your yacht to the shore. This could be in the form of a ride on a water taxi, or you could ask fellow boaters or fishermen for a lift. Most skippers won't leave their tender for a longer period tied to anything apart from a dock operated by the authorities or a yacht club.

Burglars and thieves target the yachts themselves much less frequently than the tenders. However, the vessel might be at risk if it's left unmanned for one or several nights. As a result, you should always lock the boat tightly while you are away.

Some skippers take all valuables with them and leave their main hatch open. The motivation behind this is that it allows a burglar to get inside the boat without having to cause damage by prying open any doors or hatches. While this might be suitable for crews on sparingly equipped day sailors, since they can carry all of their equipment with them, it is unfeasible for most long-range cruisers.

Despite the fact that sailors often debate the pros and cons of allowing easy access versus risking damage to hatches or frames, it is recommended that you employ strong and conspicuous-looking mechanisms. Supplemented with warning signs and alarm systems, they are quite effective at discouraging a crime before it is seriously attempted. Most burglars will try to pry a door open for no longer than five minutes in all but entirely deserted places. A good lock might very well last that long.

Getting in contact with any neighbours in the anchorage can also help. On many occasions, the crew's excursion plans are sequenced so that one crew stays on board to watch other boats while their crew members are ashore.

A local guard on your boat is another option if you do not want to solely rely on your locks or neighbours. Your prime sources when researching areas and anchorages will most likely indicate whether this strategy is viable in a specific region and perhaps list some recommended guards. In regions where excursions for yacht crews are a relevant business, tour operators can also offer guard service as part of the package. The guards will usually stay and sleep in the cockpit of your boat while the interior is locked. Prices vary widely across the various cruising grounds but usually are not prohibitively expensive in less-developed parts of the world.

Visits by the authorities: 'It's the police!' Is it really?

Visits by the authorities are recurring events during any long-term cruise. Customs, Coast Guard, Immigration or Navy – everybody wants to see your yacht once in a while, especially during the process of clearing in or out.

Unfortunately, there have been reports of boats that were approached by criminals pretending to be the police. This may involve criminals disguising themselves by wearing outfits that resemble local police uniforms – as was reported by the victims of an abduction by terrorists in the southern Philippines. In another case, the offenders approached in an unmarked vessel and simply stated they were the police. No ID, no uniform.

The fact that authorities in many developing countries sometimes do not look like a typical Western police officer makes things more difficult for visitors. One cruiser reports that he has been approached by 'ragged figures in a desolate vessel wielding Soviet-era assault rifles' claiming to be the police and demanding permission to board in the Red Sea. The skipper initially refused to let them on board but in the end he had to yield. Luckily, they actually were officials.

We had similar experiences when on one occasion the Moroccan police approached us wearing very comfortable jogging pants and flip-flops (this was during the check-in, so there was no confusion). Anchored close to Jayapura another time, three Indonesian officers approached us in a borrowed local runabout wearing civilian clothes. However, at least one of them carried a badge.

After such incidents, several authorities commented on the cases. You can extract two general rules from their combined statements:

- Usually, authorities will not attempt to board a foreign cruising vessel at night.
- During any visit, they will have some kind of ID or approach in a marked vessel.

Rare exceptions are yachts arriving at an anchorage after dark. The chances of a nightly visit may rise in areas where there's a lot of smuggling or that is close to sensitive infrastructure such as military assets.

All of this means that unless you have just dropped anchor late in the day in sensitive areas, nightly visitors claiming to be the police have to be treated with utmost caution.

What to do?

Your first action should be to make the situation as public as reasonably possible. Switch on all your lights before you start any conversation. Take a mobile VHF unit with you and covertly carry your pepper spray in case the situation turns nasty.

You could also consider calling a real or imagined friend on channel 16 while approaching the officers to inform the world about your visitors. See how they react.

Look for IDs and markings. Authorities will never set foot on your boat without asking for permission (SWAT missions excepted). If they attempt to do so, command that they wait until you give permission. If they refuse, they are most likely not real officers.

Then ask to see their ID. Although it is hard to validate a police ID in a country you are visiting, it is a good start. Maybe they do not have any.

Tell them that you are confused and worried about the visit and ask for confirmation from the police post or headquarters. An elegant way to check is to very kindly ask the officer to provide you with a telephone number for a confirmation from his superiors. Having access to a satellite phone or cell phone with a local SIM card is very helpful is this situation.

Any police officer should be able to produce contact details or try to find a solution, whereas criminals would not. Depending on their information, call the contact on your VHF, a mobile or a satellite phone. You could also ask the officer to make the connection with his phone.

If the visitors cannot be identified as officials, refuse the boarding and tell them to return during the day. If they try to board without ID, can't facilitate a confirmation call, arrive in a vessel lacking official markings and don't have permission, you have to assume that they are criminals in disguise.

Now you should start to 'raise general hell' as described previously and prepare for fending off the attackers. If you are wrong, OK, it's a bit embarrassing, but not a problem, and better than losing the initiative altogether.

SAILING IN PIRATE-PLAGUED WATERS

Only eight per cent of all 258 attacks analysed for this book were experienced at sea, and most of these true acts of piracy happened in waters well known for their dangers.

Depending on the area visited, attackers ranged from knife-wielding fishermen hoping for some extra cash to heavily armed terrorists who were well coordinated and well trained.

These regions should only be visited by crews who have agreed how much of a risk each member is willing to take and who have prepared mentally and physically for an incident of piracy.

Such preparation focuses on avoiding contact to begin with, and covers tactics for an unrelenting lookout, deception, deterrence and confrontation as well as how the crew should act after they have been controlled and how to prepare for rescue by naval forces.

> Less-professional criminals have been successfully fooled or fended off by determined cruising crews who had employed effective tactics. However, resolute and trained pirates, terrorists or parties of war will most likely succeed in subduing or killing a cruising yacht's crew.

Compared to theft, burglary and robberies at anchor, attacks at sea are very rare events. Of more than 250 crimes against yachts analysed between 2010 and 2017, only 19 were committed against vessels in transit (without counting suspicious approaches). All of these took place in areas notorious for their unstable security environment.

Such attacks can occur in coastal regions or offshore, although suspicious approaches occurring more than 100NM away from any coastline are an exception and are limited to very few regions of the world. If attacked in these areas, crews will most likely face formidable professional criminals or terrorists. Pirates prowling the coastal waters of Risk Level III regions will probably be more inexperienced or even opportunistic offenders. Successfully defending a normally equipped and crewed yacht against professionals will be very hard and dangerous. By contrast, unprepared yet courageous crews have effectively and repeatedly fended off criminal opportunists – which made up a larger proportion of attackers – in the past.

Once again, we strongly recommend avoiding any cruise in waters that have been confirmed as being troubled by professional pirates, either at present or recently. The same is true for areas affected by civil war or war-like conflicts. You will find these regions classified

FIG 4.10: *International forces approach a suspicious boat in the Gulf of Aden. (Photo: Creative Commons)*

as Risk Level IV. At the time of publishing, these included the Gulf of Guinea, a 565-kilometre (350-mile) radius around Jolo and Basilan in the southern Philippines, the coast of Venezuela and the coast of Somalia. Risky areas to be sailed only with utmost vigilance are the high-risk areas (HRA) in the Indian Ocean (including the Gulf of Aden and the southern Red Sea), the coast of Bangladesh and the coast of Honduras.

In the event that a cruise in a higher-risk-level area cannot be avoided, adequate equipment and preparation can help to reduce the chance of a successful attack on the yacht. You will find a detailed description and rating of useful equipment in Part III. Proper preparation, prevention strategies and tactics to fend off an actual attack are also described here.

Bear in mind that the sum of all measures introduced are aimed at defending your vessel against serious attackers. Sound research into the area you plan to visit helps to guide your behaviour against other vessels approaching you. The extreme caution that might be appropriate in the Gulf of Guinea, the northern Indian Ocean (Gulf of Aden HRA), Venezuela, Honduras and Guatemala, for instance, could prove overly strenuous for your crew in other regions. To meet approaching boats with a friendly yet firm and determined demeanor outside of high-risk areas is most likely the better way to enjoy your voyages.

The attackers

Attacks on yachts at sea are committed by the full criminal spectrum: professional pirates, terrorists and rather inexperienced, opportunistic offenders.

Professionals who attack yachts usually operate in notorious pirate areas, such as the coasts of Somalia, the Gulf of Guinea and the coasts of Venezuela, although there are other regions where yachts are approached while at sea. However, in those waters crews will most likely encounter the less-dangerous, inexperienced, opportunistic attacker. At the time of writing, several yachts were attacked along the coasts of Honduras, Guatemala, Nicaragua and in some areas on the Pacific coast of Colombia.

Opportunistic offenders in these areas are usually fishermen or less-fortunate individuals in coastal societies who sense an opportunity for supplementary income when they spot a yacht in their waters. Their boats are not specifically equipped for a confrontation at sea. They are manned for their original purpose and have one, two and very seldom more than three crew on board. Depending on the region, some may be carrying firearms. This is a problem in regions where there's a widespread availability of weapons, such as Latin America and the northern coasts of South America. In past attacks, the weapons used were revolvers and pistols without long-range capabilities. It might come as a surprise to many Western cruisers that carrying an AK-47 assault rifle for self-defence is quite commonplace for fishermen in the northern Indian Ocean.

When the opportunity arises, a situation that may have started as an offshore transaction for fresh fish could end up as an armed robbery. The approach of such offenders is usually inconspicuous. Sometimes they will offer fish, more often they approach requesting water, fuel or cigarettes. The decision to strike may not yet have formed in the attacker's mind. Occasionally, the change of plan occurs once they get closer and realise that the crew is small and could easily be overpowered. In other cases, the decision to attack may be made during the transaction when one crew member leaves the deck to fetch money, water, fuel or cigarettes.

Inexperienced offenders usually aim to get hold of cash, jewellery, electronics, alcohol or other drugs. Inexperienced or not, these offenders can become very violent while or after overpowering the crew.

In contrast to the opportunists, professional pirates are well prepared for an attack at sea, with vessels that are optimised for longer periods of loitering. They will carry a lot of fuel and are often equipped with powerful motors. Some may look like local fishing boats but can be identified by exceptionally large crews, sparse fishing gear and an abundance of fuel containers on deck. Professional pirates will carry some form of firearm. In the past, those reported along the coasts of Venezuela used short-range handguns, whereas their colleagues in the Gulf of Guinea and in the northern Indian Ocean were much more heavily armed. Assault rifles, rocket-propelled grenades ('bazookas') and heavy machine guns are the preferred tools of the trade in these regions.

Pirates will use ambush tactics, camouflage and high-speed, direct approaches to keep the time between detection and confrontation short. Shots to underline their determination are frequently fired from a distance.

These offenders set their sights on cash and all kinds of valuables. Only the most organised groups, such as the ones in the Gulf of Guinea, along the coasts of Mindanao and – until recently – the ones in the Gulf of Aden, will try to abduct the crew or hijack the vessel.

Terrorists and pirates are very similar in terms of tactics, equipment and lack of scruples. Terrorists usually aim at abducting the crew in order to negotiate high amounts of ransom from the crew's companies or governments. The risk of fatal violence is highest when being attacked by terrorists, since a crew killed in the action also serves their goals: media coverage and official attention for them and their ideas.

Resisting well-armed pirates and terrorists is extremely dangerous and should only be attempted when the crew is well armed, well trained or desperate not to be controlled. The risk of being shot and killed is very high in such situations.

Pre-cruise preparation

Research and risk assessment

Many sailors avoid mostly safe passages due to horror stories and hearsay that persists in the cruisers' community or the media. Such regions recently included the Strait of Malacca, the Solomon Islands, parts of Papua New Guinea and the northern Indian Ocean.

At the same time, thousands of charter cruisers and liveaboard crews roam regions that are notoriously dangerous, with many robbed, injured and even murdered victims every season. Just because they are part of the larger Caribbean cruising grounds, the dangers lurking in Venezuela, Trinidad, Latin America, St Vincent and St Lucia are underestimated.

For this reason, thorough research is of great value before cruising potentially risky waters. As a responsible skipper, you should spend a couple of hours checking out the specific situation in areas categorised as Risk Level III or IV before you actually embark. This may lead to positive surprises in some regions or help you avoid unexpected confrontations in others.

Theoretically, the risk for a specific region can be estimated by comparing the number of passages with the number and severity of attacks in the last year. Unfortunately, it is sometimes hard to get hold of the exact number of passages in a region. However, attacks on yachts as well as the tactics and level of violence employed are well documented for most regions visited by pleasure vessels. You will also get a feeling for whether the attacks were carried out by opportunists or experienced pirates. Where there have been mostly opportunistic attacks, maybe without the use of firearms, you might consider a vigilant passage. If professionals roam the area unhindered by the police and coast guard, you should definitely avoid the route.

> 'Like so many other sailors who planned to travel from Asia to the Mediterranean, we considered avoiding the convenient Suez route because of piracy threats. Quick but thorough research revealed that there hadn't been any attacks on yachts in the last 24 months. So we set sail to Djibouti without regretting it today.'
> Ankie, SY *Tinkerbel*

Your research will also uncover recent shifts in the criminals' tactics. The change of strategy used by Nigerian pirates in 2015 serves as a good example: due to decreasing prices in crude oil, they have switched from hijacking and selling loaded tankers to abducting the crews for ransom. This shift made a high-risk area even more dangerous for pleasure vessels since yacht crews suddenly became attractive targets for these criminals.

The collected information will serve as a sound basis for your decision on whether or not to take the route at all and, if you do choose to embark, how to prepare both boat and crew.

At the risk of seeming repetitive, it is wise to remember that areas where there have been reports of active, well-organised pirates or terrorists within 12 months of your planned passage should be avoided even by the most courageous of all skippers. The same holds true for areas of war and civil war. An assault by these types of offenders is almost impossible to repel by a yacht with a civilian crew and legally obtainable equipment.

Once you have decided to go for the passage through a high-risk area, you should check the local emergency and rescue infrastructure. Large navy units operate in some

FIG 4.11: *Suspected pirates are checked by armed forces. (Photo: Creative Commons)*

of these areas, others are regularly patrolled by well-equipped and organised coast guard organisations. However, many of the high-risk areas are not patrolled by any official authorities at all, leaving you pretty much on your own. Whatever the situation may be, research all possible contact numbers, email addresses and radio channels. International military organisations, the coast guard or the police may feel responsible for your target area. Do not forget to check for any proposed official or unofficial best practices for the region.

Last but not least, a check of the current and the generally prevailing weather situation is an important input for your routing through the area (see the 'Routing' section on pages 136–8 for more details).

Self-concept and strategy

A crew that has decided to sail through a high-risk area should give some thought to their self-concept, the individually perceived risk level and the general passage strategy. Most relevant topics can be addressed by answering a simple set of questions:

Self-concept

- Are there enough crew to keep a continuous lookout?
- Are we physically fit enough to repel a boarder at sea?
- Are we willing to hurt other humans if we are attacked?
- Are we properly equipped to avoid and repel an assault?
- Do we prefer to risk health and life in a confrontation or do we prefer to take the consequences of being controlled by criminals (e.g. loss of valuables, injuries, abductions, mock executions and possibly death)?

Basic strategic thoughts

- How will we react if we spot a suspicious vessel?
- How will we react when a suspicious vessel approaches? Determined deterrence at the earliest stage or passive recon until the vessel reveals its intentions?
- Are we willing to invest a bit of time in preparing for the worst case?
- At what point will we surrender? If there's an offensive approach? When shots are fired? When one of us is injured? Not before boarders are on deck?

> *'Prior to setting sail from Sri Lanka to Sudan, we needed to decide whether we would surrender or resist if there were an assault by expected opportunistic offenders. After researching many cases of piracy and the specifics of our cruising ground we decided to resist aggressively rather than being controlled. We equipped our boat and prepared our family crew accordingly.'*
> Heide, SY *Alytes*

Every crew member should still feel confident after answering the first set of questions. The crew must also agree on a basic strategy drawn up by answering the second set of questions. Anything but a consensus on these matters should lead to the abortion of the passage in the high-risk area and to finding an alternative. Dissenting views on the fundamentals such as risk perception and behaviour in a confrontation will increase the risk if there is an attack. It will most likely lead to a generally more stressful and unpleasant cruise, too.

Routing

Routing is one of the two most important security factors when cruising a high-risk area (an excellent lookout being the other).

When plotting your course, you should aim to minimise the time you'll spend in the area and to avoid all of the regions in which there is an increased probability of being spotted by pirates. A secondary goal – partly coinciding with being stealthy – is bypassing any obvious ambush sites.

The rationale for your first goal – minimising time spent in the area – is simple: every day you are not spending in pirate-infested waters reduces your risk of being spotted and attacked. A sailboat skipper should understand that achieving this means that you can't always take the shortest route through the area. You will need to consider weather and your vessel's performance to define the course that minimises time. To add to the challenge, bear in mind that you simultaneously also want to avoid ambush sites and the probability of being spotted.

As you should have researched the area at this point, you will have some information about the pirates' capabilities and strategies. Now it will be helpful to know how far they usually venture out to sea from their coastal home bases. The distances travelled by pirates vary widely: less than 100NM in Latin and South America, more than 300NM for groups operating in the southern Philippines and up to and over thousands of nautical miles for the pirates who roam the northern Indian Ocean. Ensure that your information about pirate behaviour is as up-to-date as possible since some groups may have ceased to exist while others might have improved or changed their strategies over time.

Knowing the typical maximal range of the offenders in your cruising area helps you to maintain a distance that should prevent you from meeting them offshore by accident. What's more, you should generally keep out of sight of any coasts in the area to avoid being spotted and targeted from there. Generally speaking, the further away you are from any coast the better. Even if it means more time in an area: keep away from the coasts.

If geography or weather forces you closer to a coastline, try to stay out of sight. Pirates will simply not be aware of you unless they are capable of using radar or employ sentinels at sea that carry radios (both strategies have not been reported, so far).

In order to get an estimate on how far away from the coast you should stay, you can use a very simple formula and a calculator. You need to include the height of your vessel as well as the expected height of a suspected spotter's perch:

First, enter the values in metres. The result will be an estimate in kilometres. For a value in nautical miles, you have to divide by 1.853 in a last step. When trying to estimate the height of a potential spotter you should take into account that he will most likely not be sitting on the ground at sea level. Taking into account the coastal topography close to the beach, it will be safe to assume an additional 7m. This is roughly the roof height of a two-storey house.

As an example: if your chart shows a height of 10m close to the tide line, add 7m and

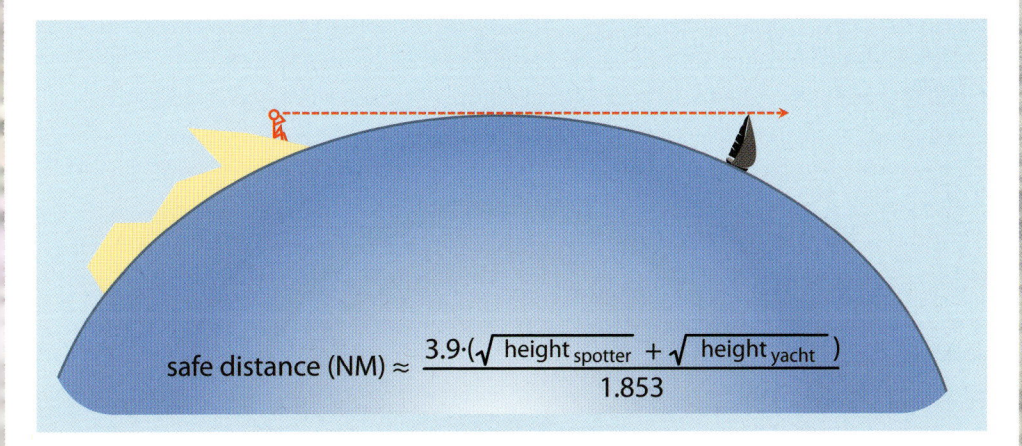

$$\text{safe distance (NM)} \approx \frac{3.9 \cdot (\sqrt{\text{height}_{\text{spotter}}} + \sqrt{\text{height}_{\text{yacht}}})}{1.853}$$

FIG 4.12: *Estimating the safe distance to stay from shore by gauging the height of the spotter and the yacht.*

insert the resulting 17m into your formula. If you now insert the height of your mast – let's put it at 20m for this example – the resulting safe distance to shore is at least 18NM in very clear weather. If you assume a spotter is sitting 100m high up on a cliff, your safe distance increases to 31NM. It is a good idea to check your charts for hills, towers and chimneys and add their height to the spotter's height.

Bad weather and atmospheric effects will reduce theoretical visibility to your favour, in most circumstances. Even in very clear sky conditions, a typical sailing yacht cannot be spotted at distances greater than 27NM.

Apart from being stealthy, keeping a distance of at least 12NM from any coast will also help you to avoid misunderstandings with local authorities and the risk of being deceived by pirates who have made their approach disguised as police or the coast guard. Fending off bona fide officials – who are simply hard to identify because they lack uniforms and official insignia – with machetes and pepper spray may result in a lengthy sojourn in a rather unpleasant prison. However, outside of the 12NM-zone you can be sure that everyone who approaches dressed as the police or the coast guard is probably trying to deceive you.

When plotting your course, try to avoid passages between islands of any archipelago because these are excellent ambush points that put the targeted yacht at a serious disadvantage. There have been several reports of vessels being assaulted or at least being shot at in these settings (the last report describes a yacht being fired at when approaching the Hanish Islands in the Red Sea).

If a passage close to the coast or amush sites cannot be avoided, the crew needs to keep a sharp lookout and be fully prepared to repel boarders until they can once more move a safe distance from the shore.

Preparing the yacht

A yacht that is planning to sail through waters plagued by pirates must be in the best possible shape. Without venturing into the realms of good seamanship and technical preparation for any passage on a yacht, we do want to stress that all systems needed for propulsion, navigation and communication must be reliably working for the duration of the estimated time of the passage. As already described, every day in dangerous waters increases the risk of an assault. You will therefore want to avoid any delay caused by having to make repairs at sea. To know that your engine will start and run without the slightest doubt is a reassuring feeling should you need engine support for an escape route or an offensive manoeuvre towards a suspicious vessel.

It has proved extremely useful to store in your satellite phone's phone book and speed dial the telephone numbers of the local (or your home countries) MRCC, the coast guard and any other authorities (military or police) operating in the area. Make sure that all crew members know how to operate the phone.

To have some contact details (phone numbers and email addresses) of your EPIRB's registering authority at hand may come in handy if you trigger an alert that you want to revoke after the situation has been resolved.

A printout summarising all relevant contacts as well as the most basic steps for calling for help will enable even less-experienced crew members to effectively support emergency communication should there be an assault (see appendix for a draft).

Useful security equipment, from improvised fender barriers to firearms, is described and rated thoroughly in Part III of this book.

Before setting sail, you should install or store in useful positions alarm systems, emergency signals (flare guns, parachute flares etc) and possibly weapons. All alarm devices and emergency signals are best kept close to the lookout's station. This means that crew or watch will not be forced to leave their station to warn the crew or initiate an early engagement with an approaching offensive vessel. Whenever the crew is below deck, the weapons should be stored nearby for easy access in case the alarm is triggered. Not storing weapons on deck also ensures that the crew will still have access to them in situations in which attackers have surprised the lookout and are already aboard.

During these final preparations, the crew should also install any boarding barriers and any tools that are required to ready the vessel for police or military intervention.

You may also consider preparing for the possibility of a successful boarding. According to international military forces, you can support a military or police rescue by disabling the boat. This needs to be done quickly the moment you think you will not be able to repel any professional pirates who you presume are planning to hijack the vessel and the crew. Think about placing a reliable blade close to the mast that you can use to sever halyards and sheets, a pair of pliers close to the engine starter batteries, some items to block the rudders and companionways, and some tape to block the VHF radio's transmit button on channel 16 for an easy triangulation. For details of these preparations, see 'Rescue by military and police units', page 160.

Preparing the crew

A well-prepared crew is a critical factor in successfully deterring offensive approaches and effectively fending off any assaults on your vessel. A trained and confident crew will also be the basis of a relaxed and generally successful passage: every person aboard will feel reassured the moment they are certain that they are not only sharing a common strategy but also know their (and everyone else's) role should there be an assault.

To prepare a common strategy, make sure that the essential questions in the 'Self-concept and strategy' section (see page 135) have been addressed and decided on by skipper and crew. To prepare an effective defence in case there's an actual assault on the vessel you will need to familiarise your crew with and, at least briefly, train them in the tactics described in 'Crew tactics: avoidance and defence' (see page 145). Every crew member should know the general sequence and every aspect of deterring and defending the vessel.

This will include the use of emergency signals, communication equipment, defensive equipment, weapons and replicas as well as defensive positions and the movement patterns between them. Familiarising the crew with basic defensive tactics does not require many days of training and, however awkward it may seem for a civilian crew, it definitely pays off to play through the sequences as often as necessary to have everybody on board comfortable with it.

The dynamics of offshore attacks

As a crew will always have to be overpowered for an offshore attack to be successful, these crimes are categorised as 'Controlling Attacks.' They roughly follow the same dynamics as all offences in these groups.

The classic assault at sea follows the sequence 'Ambush', 'Approach', 'Board' and 'Control'. Attackers will try to maintain the element of surprise for as long as possible throughout the attack. Sneaking up on an unprepared boat and overpowering the sleeping crew (not unrealistic on a liveaboard vessel) would be the best course of action from the pirate's perspective.

The most important weapon in a conflict with pirates at sea is continuous vigilance. The moment a yacht's crew succeeds in spoiling the element of surprise, their chances for successfully averting an attack grow substantially. Have a look at the dynamics from both perspectives.

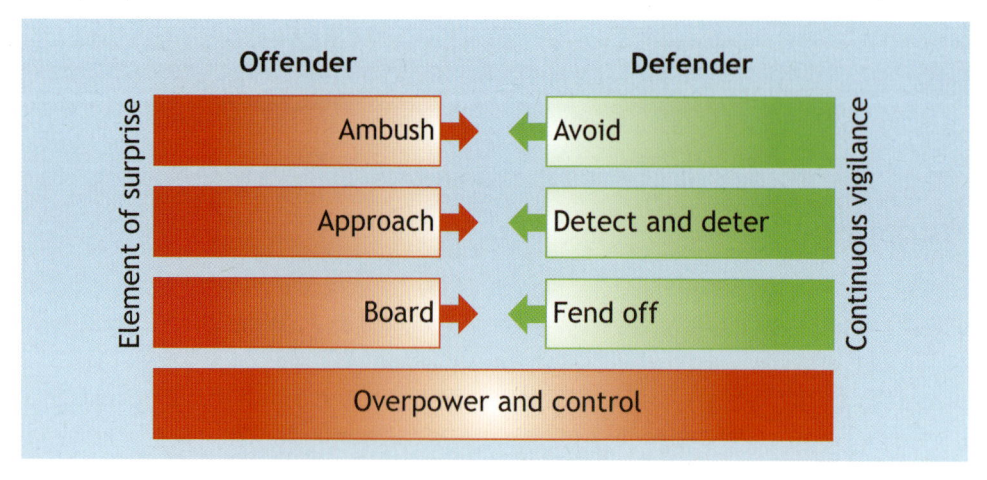

FIG 4.13: *Dynamics of a pirate attack – surprise vs vigilance.*

Pirate tactics

The attackers will attempt to take control of your vessel while minimising their overall risks. To do so, they will try to hide themselves or at least their true intentions through the first three stages of the attack. This strategy of stealth primarily aims at reducing the time that defenders have to resist as well as minimising the number of crew that have to be overpowered at the time of boarding.

Depending on their resources, pirates' tactics to keep the crew in the dark vary. Most opportunistic and less-experienced offenders will rely on camouflage and mimicry, while professional pirates are known to employ speed and violence of action during their approach.

Ambush

The ambush's overall goal is to minimise the time that a yacht crew may have to prepare for defensive actions and to make emergency communication. The attackers therefore aim to take up a position as close as possible to the targeted yacht without being noticed or at least without revealing their malicious intentions. They will try to reduce the time of approach across an open body of water as much as possible. Techniques of stealth, camouflage and deceit are typically used.

◀ Yacht   ◀ Attacker

FIG 4.14: *Local topography offers ambush sites along a disadvantageous (for the defender) coastline.*

In coastal waters, pirates will use the topography to remain hidden for as long as possible as a yacht approaches. Coastlines featuring bays, crevices and fjords are especially useful to pirates in this respect. Archipelagos are even more advantageous: they not only offer excellent hiding places but limit a target yacht's manoeuvrability and performance, too.

When offenders can't make use of the benefits of a rugged coastline for a good ambush, they have instead successfully used deception in the past to get dangerously

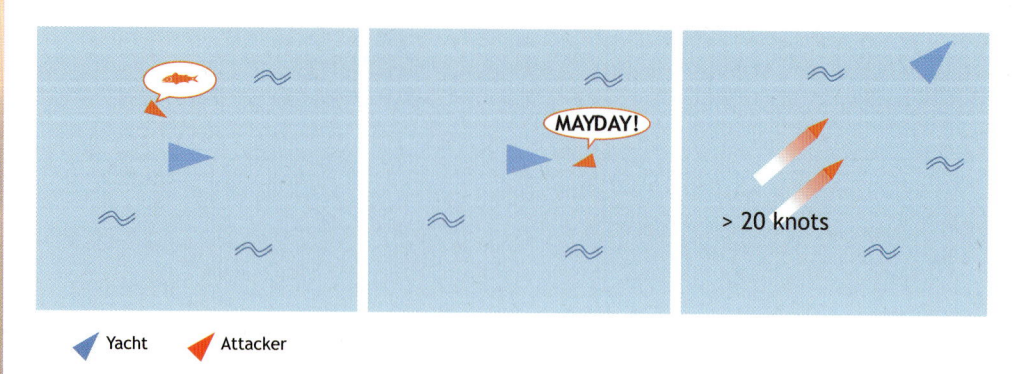

FIG 4.15: *Pirates like to use deception and speed to get close to a victim at sea.*

close to an unsuspecting yacht. Pleasure vessels have been approached by attackers camouflaged as fishermen, with requests for fuel, water or cigarettes or even lured close with fake emergencies.

Pleasure vessel crews have proven to be fair game for clever deceits in several offshore attacks in the past. Helping a fellow mariner in distress or wanting to buy fresh fish from friendly locals are normal, expected behaviours. However, in these cases, the moment the pleasure craft is alongside, attackers only need to draw a concealed gun to effectively control the crew who are lined up expectantly at the guardrail.

It is quite surprising that offshore attacks rarely occur under cover of darkness, although perhaps the risks involved when boarding and getting your bearings on an unknown vessel at night may be the cause. It is also harder to support a boarding party from the attacking vessel at night, since most weapons used by pirates cannot be aimed effectively in the dark (and may simply not be seen by the targets, which reduces the threatening impact of their appearance). However, the general trend for daylight assaults may shift as night vision equipment becomes more available to marine criminals, so it can't be ruled out.

In contrast, pre-dawn or dawn assaults are quite common. The first pre-dawn light eradicates many disadvantages for the attackers, while offering the considerable advantage that the majority of the crew is likely to be unprepared and sleeping in their cots.

Offensive approach

Aside from a successful deception, the offensive approach is an element of most pirate attacks. Depending on the pirates' capabilities and their success in getting close to the target, this will occur from a great distance 'over the horizon' or from a more nearby position.

Whenever any unknown vessel approaches your boat fast and in a straight line to a distance closer than 1NM you should define it as an offensive approach. The same holds true for a passing manoeuvre on the open sea.

If weapons and/or boarding equipment (such as large ladders) are visible or being

FIG 4.16: *A boat carrying a boat and a lot of fuel drums in the Gulf of Aden. The people on watch who took this photo have to be vigilant and keep their cool – it could be a cargo barge or smugglers – neither of which are a problem, as long as they stay away. (Photo: Berswordt)*

prepared, you have to brace for an imminent attack.

This rather conservative definition of an offensive approach could be a problem in some regions. Harmless fishermen, ambitious traders or nosy coastal dwellers may close in on your vessel in even the most notorious regions. Not all people living in such areas are aware of the danger that a misinterpretation of their behaviour could pose should the crew aboard the vessel they approach be armed. The risk of approaches that are falsely categorised as being 'offensive' increases rapidly the closer you get to a coast. Tactics for how to cope with that situation without endangering harmless mariners are described in the section on 'Crew tactics' (see page 145).

From the attacker's perspective, the perfect situation is when an approach goes unnoticed until the moment the vessel is alongside (due to a lack of or careless lookout on the target boat). However, every second and every metre the pirates gain on their target without giving the crew a chance to prepare will shift the odds in favour the attackers (due successful deception by the pirates or lack of preparation by the target crew).

Most professional pirates will close in at maximum speed (> 20 knots) with at least two skiffs from multiple directions. The skiffs will each be manned with no fewer than three people. Offenders operating along the coasts of Somalia and Nigeria will present their weapons during the attack. They usually carry assault rifles and rocket-propelled grenades, which pose a great risk for most yachts and their crews. In the Gulf of Guinea heavier weapons such as large-calibre, deck-mounted machine guns have been reported. It is not uncommon for professional pirates to fire shots at the target vessel to underline their determination to make it stop so that they can board. In most cases, these first signs of aggression begin at a distance of 300 metres (yards).

Pirates outside these notorious areas are currently less well armed. The use of an assault rifle was only reported in one attack (close to Venezuela) over the last five years. Nevertheless, firearms such as revolvers or pistols were used in some offshore attacks. To compensate for the missing long-range capabilities of their weapons, criminals in most parts of the world will try to get close to your yacht and overpower your crew from the side of your boat. Though firearms are extremely dangerous, you should bear in mind that effectively aiming a short-barrelled handgun in a marine environment between two moving vessels is extremely difficult. Consequently, point-blank is really the only range pirates can use to cause harm with any more likelihood of success than mere chance.

Boarding

For a successful attack at sea, the pirates need to board. This mission-critical manoeuvre poses some risks for the offender. This is because it is the last phase during which a crew determined to defend their vessel can prevent the attackers from taking control before they have to expose themselves to the uncertainty of battling the offenders on the yacht.

Depending on the size of the attacking force, either all of the pirates will try to board or some will remain in the stable safety of their vessel using firearms to threaten the yacht's crew from there.

Pirates seem to fundamentally differ in their willingness to take risks, and the defender's strategy may be adapted accordingly. Opportunistic criminals in the Caribbean and along the coasts of Latin America are more reluctant than those prowling the Gulf of Guinea, for instance. The latter are well known to engage onboard security teams in lengthy gun battles. These are risks that even Somali pirates were averse to taking. Instead, they would retreat after detecting armed guards or after warning shots had been fired.

Any attacker will try to set the stage so that he can board a motionless vessel with a crew already overpowered or controlled from a distance. In this scenario, the crew sits or lies on deck, disarmed and hands visible. In the worst case, they assist the offenders in stepping over to their vessel.

To step over to a yacht in motion, even in a moderate sea state, is a risky manoeuvre for any boarder, especially if the crew is resisting either overtly or from covered positions (depending on both the attackers' and the defenders' armament). The need to use their hands while stepping over makes them quite vulnerable at the moment of boarding and there are several reports of yacht crews who successfully managed to fend off boarding attempts by poorly armed pirates from rather slow boats.

Establishing control

Control needs to be established and maintained. The first stage, the actual overpowering of a crew, will at least involve threats of violence and death. In many cases, the offenders will commit acts of violence varying from a mere shove to the lethal use of guns against some or all crew members.

Maintaining control is easier than establishing it since victims will probably be physically bound and mentally broken.

In our recent study, about half of the crews who did not resist in any stage of a robbery suffered injuries of some kind or another. Some were inflicted during an overly aggressive overpowering of an already submissive group of victims, others while attackers forcefully questioned the crew about money stashes or simply because the attackers were particularly aggressive. We can only speculate about whether or not the abuse of stimulating drugs was a contributing factor to such behaviour.

Depending on the attackers' plans, control will only be maintained until boat and crew are stripped of their valuables, or for prolonged periods in cases of kidnapping.

When taken hostage, the groups involved will try to transport the victims into a region where there's a suitable supporting infrastructure for the offenders. The subsequent imprisonment is reported as being very unpleasant, no matter which group has commandeered the crew. To keep a group of people subdued for such a long time involves the application of physical restraints, starvation, general hardship, constant threats, mock executions and general acts of domination. All of these activities result in utmost discomfort for the victims.

Crew tactics: avoidance and defence

The good news: there are relatively numerous reports of crews who have effectively fended off a suspicious approach or a definite act of piracy. While attacks by professional pirates have so far not successfully been repelled by non-military personnel, yacht crews *have* managed to defend their vessel against less-experienced or opportunistic pirates.

The most important techniques to use to avoid or fend off an attack are a continuous lookout, determined deterrence and resolute repelling. How you can employ these before and during an actual attack is described in the following sections, ordered according to the phases of an attack (ambush, approach, boarding and control).

However, you need to remember that innocent mariners without evil intent may approach you even in the most troubled waters. You and your crew have to give each vessel the chance and the time to change their course as long as they are not unmistakably identified as being hostile. The reaction scheme below will help you to open up that opportunity for an approaching boat without sacrificing any of the valuable time needed to stage an effective defence, if necessary.

As the general sequence of action/reaction described in the scheme ensures the safety of all innocent parties involved, your tactics must change as soon as the approaching vessel is identified as having ill intent. Any direct, fast approach leading to a range of less than 50 metres (yards) from your vessel despite all counter-actions described in the scheme having been taken can well be considered hostile. No seafarer will plot and maintain such a course without the intent to harm the targeted vessel. To eliminate any doubts, it is vital that your crew executes all warnings early, resolutely and without ambiguity. Otherwise, your 'stay-away' signals may not be understood properly. The moment the crew of the approaching vessel clearly gets ready for an attack (preparing or presenting weapons or boarding tools), your crew should immediately abort all warning efforts and initiate determined defence measures and calls for help.

According to international law, it is neither necessary to keep up warnings at this point nor would it be tactically appropriate. Any uninvited attempt to board your vessel is also clearly defined as an

Ambush-avoidance zone
- As long as no ships are visible, stay inconspicuous and keep a continuous lookout
- Upon detecting a conspicuous vessel, inform crew and prepare defensive systems
- Consider early course change to win time or force attacker to show his intent

▼ your yacht
▼ suspicious vessel

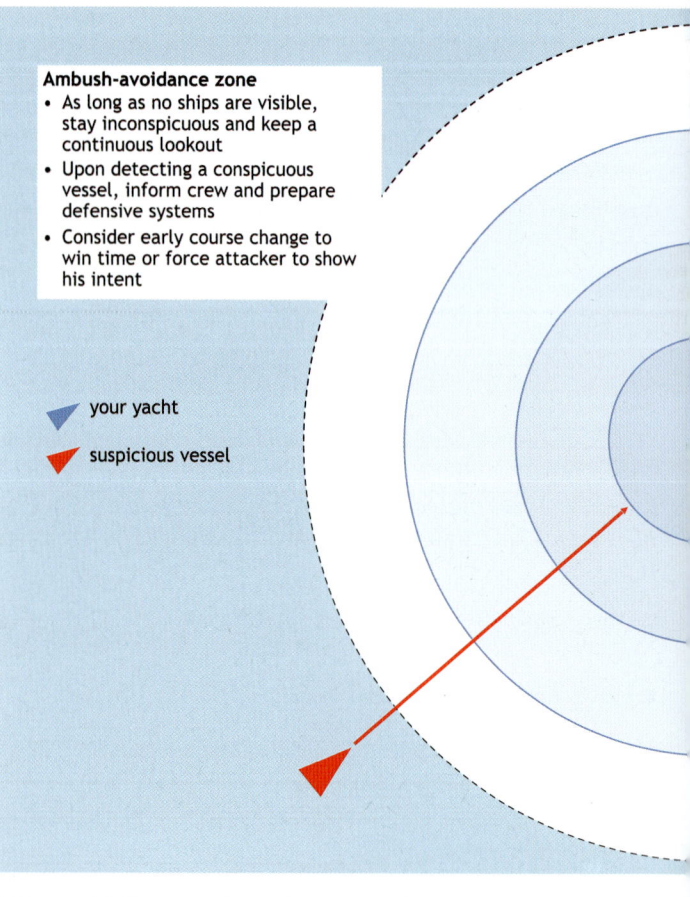

FIG 4.17: *Vigilance and reaction zones when a suspicious vessel approaches.*

act of piracy and should trigger the same change of tactics: from warning to determined fending off with all means available to you.

Tactics concerning the deployment of emergency signals also change in pirate-plagued waters: it is always better to raise the alarm sooner rather than later. To call MAYDAY and cancel it soon afterwards is perfectly acceptable in these regions. Rescue parties and coast guards will prefer to embark on one mission too many than one too few.

There is a second aspect to emergency signals. Several reports of attacks at sea state that suspiciously approaching boats and confirmed attackers have aborted their actions the moment they perceived emergency signals (sirens, MAYDAY calls and flares). There are several reasons for this: the signals clearly indicate that you feel threatened to anyone who is just being nosy, and they back off; a determined attacker will know that you are not

SECURITY IN ACTION 147

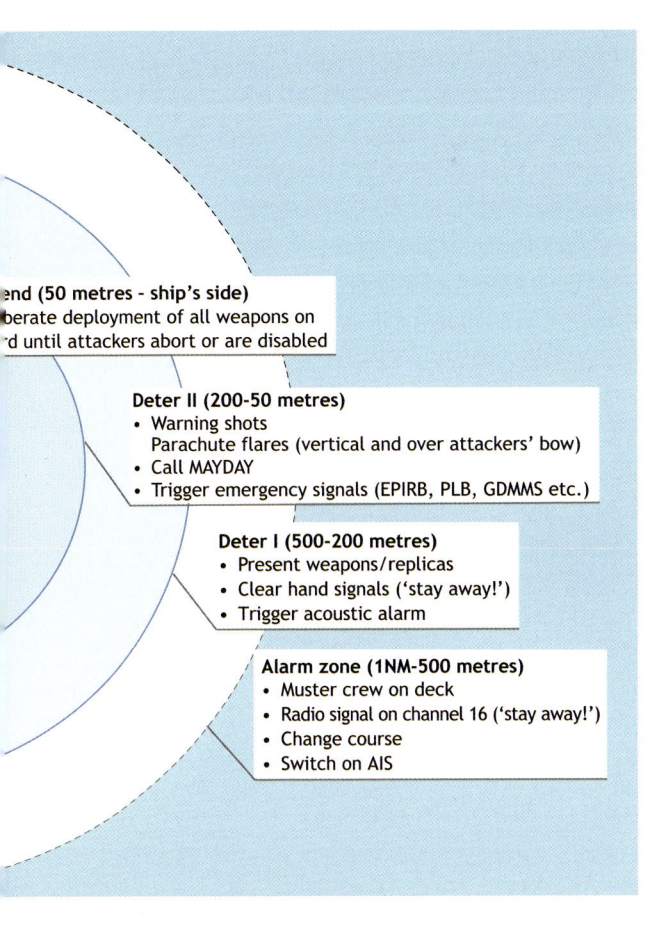

end (50 metres - ship's side)
berate deployment of all weapons on
d until attackers abort or are disabled

Deter II (200-50 metres)
- Warning shots
  Parachute flares (vertical and over attackers' bow)
- Call MAYDAY
- Trigger emergency signals (EPIRB, PLB, GDMMS etc.)

Deter I (500-200 metres)
- Present weapons/replicas
- Clear hand signals ('stay away!')
- Trigger acoustic alarm

Alarm zone (1NM-500 metres)
- Muster crew on deck
- Radio signal on channel 16 ('stay away!')
- Change course
- Switch on AIS

surprised and are wide awake; and long-range signals such as a DSC MAYDAY on the VHF radio or red parachute flares trigger uncertainty in the attackers' minds. From this moment on, they cannot be sure if and when rescuers or witnesses will arrive on the scene in boats or planes.

Remember that an unrelenting, keen lookout is the most important aspect of successful counter-piracy activities. The impact of each reaction following the detection of a suspicious vessel is very much dependent on *early* detection. The more time a crew has to issue warnings and commence deterrence-activities, the more likely it is to escape unscathed. Situations that see attackers able to close in to less than 30 metres (yards) without being detected or deterred are likely to end in catastrophe.

Let us have a look at the alarm zones for a successful deterrence.

Avoiding an ambush

A simple two-aspect strategy is enough to help you avoid an ambush by pirates: be stealthy and detect the other party before they detect (or actively engage) you.

Being stealthy and staying undetected

In the section on 'Routing' (see page 136) we described why and how to plot a course away from the coastline. This tactic is key when trying to stay undetected, so keep in mind the important rule: stay away from the coastlines. If you can't, stealthiness is hard to maintain. Consequently, when sailing close to coasts, crew and equipment should be prepared and alert the whole time; your lookout needs to be even more careful than they are when you are sailing further away from the coast.

What else can be done?

Many skippers wonder whether it is a good idea to put out all lights on board at night, and there are lively discussions about the use of radar and AIS, too.

Deciding which systems to switch off depends on the nature of the region you are sailing in. The relevant factors are the general traffic as well as the strength and capabilities of supportive units (military, coast guard, SAR) in the region. As a rule of thumb, light discipline is a very helpful practice to minimise your detectability in the night.

The 'Best Practice Manual 4' published for the Horn of Africa recommends keeping all lights switched off except basic running lights (red, green and white). Some very cautious skippers will even dispute the use of nav-lights.

Although the lights for small pleasure vessels are specced to be seen from 2NM, most of the modern units carry well to the horizon on a clear night. The higher they are installed, the further they will be seen.

However, to carry and use navigation lights is a requirement stated by the 'International Regulations for Preventing Collisions at Sea' (COLREGs). Whenever a skipper decides to switch them off, he may weaken his position in a trial or insurance negotiation if there is then a collision at night (should he survive the event). Therefore, he needs to weigh the

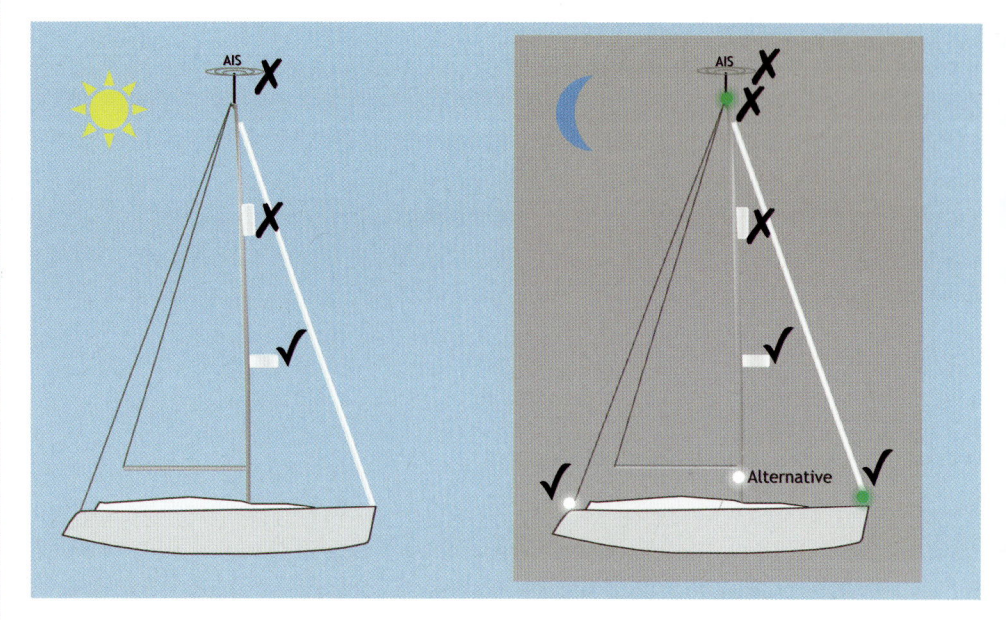

FIG 4.18: *Stealth tactics during night and day – navigation aids when running silent.*

definite stealth upsides of running black against the risks of being overlooked and rammed by an unsuspecting ship.

As an alternative, you may at least consider avoiding placing your tri-colour lantern in the masthead. As shown in Fig. 4.18, we think a low-lying running light is a useful compromise.

Given that you will maintain a continuous and keen lookout, the risk of a collision is actually quite low in any region that is not greatly frequented by commercial vessels. Freighters and tankers usually run straight courses and are easily dodged when detected early. Fishing boats, on the other hand, often navigate erratically, making them hard to avoid when you are trying not be seen yourself. However, some captains may say that most fishing boats will run you over whether or not you have your lights on!

Actually, a number of skippers have sailed in scarcely frequented waters without running any navigation lights. Indeed, before the advent of low-power LED running lights, this was a common power-saving strategy used by circumnavigators.

Crews in regions with aggressive, and sometimes armed, fishermen developed the alternative of disguising themselves as 'local boat' by just rigging a 360-degree white lantern.

Some skippers prefer a more flexible approach for reasons of safety, running their lights up to the moment they see another boat behaving strangely or on a collision course. Upon detecting such a situation, they dowse all lights and change course to evade the situation by running dark.

Regions densely populated with fast-sailing freightliners and tankers, such as the Strait of Malacca or the Gulf of Aden, call for a more cautious approach. Your navigation lights should be turned on during your passage in these areas. All other lights on board should be switched off until you are certain that you cannot evade a suspicious vessel.

The radar system is a very valuable addition to your own lookout capabilities and it should not be sacrificed out of fear that pirates are equipped and trained for effective signal intelligence. Although cheap radar detectors are readily available in Western electronics stores, as yet there have been no reports of their use by pirates. What's more, although they are effective in alarming crew to the presence and rough direction of a radar source, it would be very hard to effectively track a vessel over a longer distance with their help. This is especially true if you do not operate your radar continously.

Your radar reflector may be a bigger concern when navigating in a high-risk area. Simple radar systems are quite widely available (legally obtained or looted from other vessels) even in more remote areas of the world. De-installing your radar reflector, therefore, greatly reduces your visibility on those systems and is therefore worth a thought. As you will be on a constant lookout, the risk of being run over by a commercial ship is minimal.

Using the very popular AIS transceiver is another way of greatly compromising your safety. When your unit transmits an AIS signal, you can be detected and tracked very easily; potential attackers can follow your movements accurately in distances ranging from 5 to about 25NM around your position if they possess an AIS receiver. These systems can now be bought online very cheaply. Worse yet, pirates can at least obtain a general course and ETA of yachts heading for their area by simply checking publicly available websites.

It is therefore highly recommended that you deactivate your AIS transponder if you carry such a unit. In fact, you should actually switch them off *before* altering your course towards a high-risk area. Being a passive system and very useful for your own lookout, receivers should remain active.

NB: Most authorities recommend that you activate the AIS transmitter the moment you are approached by a suspicious vessel. As you will have already been detected, it no longer represents a disadvantage, and reactivating it will mean any potential supporting units will have an easier time locating your vessel after your call for help.

Detecting the threat: the effective lookout

The bad news first: in high-risk areas, one crew member *must* be on lookout at all times. Awake and alert. As stated before, detection of suspicious vessels at the earliest time possible is critical for avoiding and fending off any attack on your vessel at sea. The more time you have to react, the more effective your measures will be.

Consequently, the most important duty is your continuous and keen lookout. Day and night. The crew must not be caught off guard by any approaching vessel.

To get a feel for the situation on the water: if your lookout is unreliable and allows a skiff travelling at 20 knots to reach your alarm zone (1NM of distance), the pirates would need about three minutes to reach your swim platform. First shots might be fired in half that time.

If your lookout is alertly standing at the helm, his eyes will be at an approximate height of 3m (10ft) (depending on your type of vessel). The pirate skiff just introduced is of an assumed height of 1.5m (5ft) and will theoretically be detectable at a distance of 6.2NM. Assuming a speed of 20 knots, your crew will have a trifle more than 18 extremely valuable minutes to signal, call for help and prepare weapons as well as themselves for a confrontation. Each mile the skiff closes in will be three minutes lost.

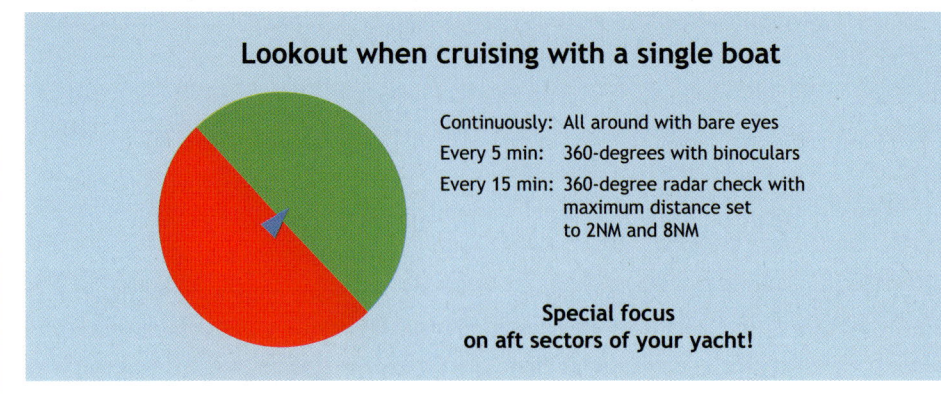

Lookout when cruising with a single boat

Continuously: All around with bare eyes
Every 5 min: 360-degrees with binoculars
Every 15 min: 360-degree radar check with maximum distance set to 2NM and 8NM

Special focus on aft sectors of your yacht!

FIG 4.19: *Lookout tactics – focus and scanning intervals in high-risk areas.*

SECURITY IN ACTION

Some crews have reportedly made a sport of being on lookout: each vessel detected outside a range of 2NM that was not shown on the vessel's AIS counted as a point. At the port of destination, the winner was treated like a star and received princely rewards.

To increase the chance of an 'at horizon detection' the lookout should scan the sea with binoculars every five minutes. Bear in mind that the usually neglected aft sector should now be in focus since attackers prefer to attack from behind.

If radar is installed, it should be used to swipe the area at least every 15 minutes. Use a resolution of 2NM between rings (8NM' maximum distance in most systems) as well as a 500m resolution (2NM' maximum distance in most systems). The latter is especially useful for spotting small, unlit vessels at night since their outboard usually shows up in this resolution. Use your radar frequently – your power management should have been planned so that radar can be operated in the recommended intervals.

If your boat features an AIS receiver, quickly compare your radar echoes with the AIS signals and focus on all vessels that do not transmit AIS.

Taking note of any contacts, with time and bearing information, in a simple notebook will help you to estimate the course and position of your target. If you identified it on radar, you should also note the distance. Many modern radar systems allow you to appoint so-called MARPA (Mini Automatic Radar Plotting Aid) targets. Depending on the model of your system, MARPA will track the signal, calculate speed and direction, and will raise a collision alert once such a situation is forecast. All of this means that MARPA is extremely helpful for staying on top of things when using radar.

If your navigation systems are able to plot a radar overlay on your GPS chart, the identification or visual spotting of any target around you really is a piece of cake.

Once detected, all targets that do not transmit AIS should be closely monitored until they have vanished over the horizon. Should they change course and come towards you, the alarm needs to be raised.

When thinking about raising an alert (or a 'heads up' if the contact is far away) the lookout should not discriminate between vessels. As already discussed in the sections on pirate tactics, offenders in the past have successfully used deception to get close to their targets. They may offer fish or ask for water, fuel or cigarettes.

For this reason, in high-risk areas the crew should not allow *any* boat closer than 1NM. The moment a vessel crosses into the 1NM-zone, your crew should initiate their warning, signalling, deterring and ultimately using defensive measures as indicated in the reaction scheme.

Apply the same caution when an apparent emergency is called in your vicinity. International maritime law rightly calls for mandatory assistance, but it also includes a moderating term stating that you only need to help if your own vessel and life are not at risk. Accordingly, never get closer than 1NM to a vessel in distress while cruising a high-risk area. Moreover, you should keep a very keen lookout in all other sectors to detect any vessels that may close in while you are concentrating on an apparent emergency. The only action to be taken in such a situation is to shadow the vessel at a safe distance and relay their MAYDAY with your satellite phone or radio to the nearest professional rescue organisation.

Only if the situation on board deteriorates clearly and visibly should you further approach a stricken vessel. By this I mean a major fire or a sinking vessel. Depending on the sea state, it may be safer to wait until the crew is in the water and rescue them from there rather than approaching a vessel that looks disabled. It is strongly suggested that you carefully search any rescued crew for weapons and other dangerous items while you can control them.

The following section will describe alarm, deterrence and defence. All equipment used to this end is discussed in the chapter 'Preparation' (see pages 48–105).

The ambush-avoidance zone

This describes the zone outside your 1NM perimeter. Whenever you detect suspicious vessels on the water around your boat, try to assess their intent (harmless or hostile).

A strong indicator for trouble is the so-called offensive approach. This is defined as any vessel that is approaching at high speed on a collision course.

If you detect an offensive approach (even outside your 1NM perimeter), you should definitely initiate all measures to alarm the crew, ready your equipment and start to deter the suspicious vessel.

The earlier, the more determined and the more offensive your crew acts, the less successful an assault on your boat will be. When reacting to an offensive approach, the crew should not do things by halves as such behaviour may lead to misunderstandings, insecurity and to advantages that favour a potential attacker. It is wiser to risk scaring a harmless fisherman by presenting a weapon (or replica) from a distance than to be surprised by a gun-wielding opportunist at your yacht's side. Warning shots from available weapons or emergency signalling devices are OK in this situation. Keep approaching vessels at a distance.

Don't hesitate to call for assistance. It is a lot better to revoke a prematurely called MAYDAY than to be boarded without anyone knowing. When in doubt, shoot that parachute flare and revoke via radio later. It will be much more beneficial than keeping both rescuers and the attacker uninformed of the fact that you feel threatened and that you are not to be taken unawares. Emergency communication is an important tool in deterring pirates. Calling for help and revoking is perfectly harmless in high-risk areas. Not deploying them would be foolish.

You might also consider an early course change if any suspicious vessel is detected approaching offensively. At this point, every course change should point away from the approaching vessel, to win time. It will also force any pirates to show their intent by altering their course to get closer to you. Any course change towards you after you just turned away is a strong indicator of ill will.

Rest assured that the attackers will not be lax in watching your vessel and constantly balancing their risks against their possible chances. The earlier they get a clear signal that they have lost the element of surprise and that their booty may come at a high price, the better the situation for the defending crew. Conspicuous defensive elements such as fences, transom barriers or warning signs installed before embarking will now be effective without requiring a single crew member's attention.

The 1NM perimeter: alarming the crew

The primary goal of this phase is to ensure an alert crew, have them ready for any measures and (if the crew is not too small) demonstrate numbers on deck. The lookout should raise the alarm in the following situations:

- A suspicious vessel is approaching on a collision course and **approaches** the 1NM perimeter.
- A suspicious vessel is **already inside** the perimeter due to a successful ambush or a careless lookout.
- The yacht plans to **close into visual range** of any coast or to pass between islands of an archipelago.

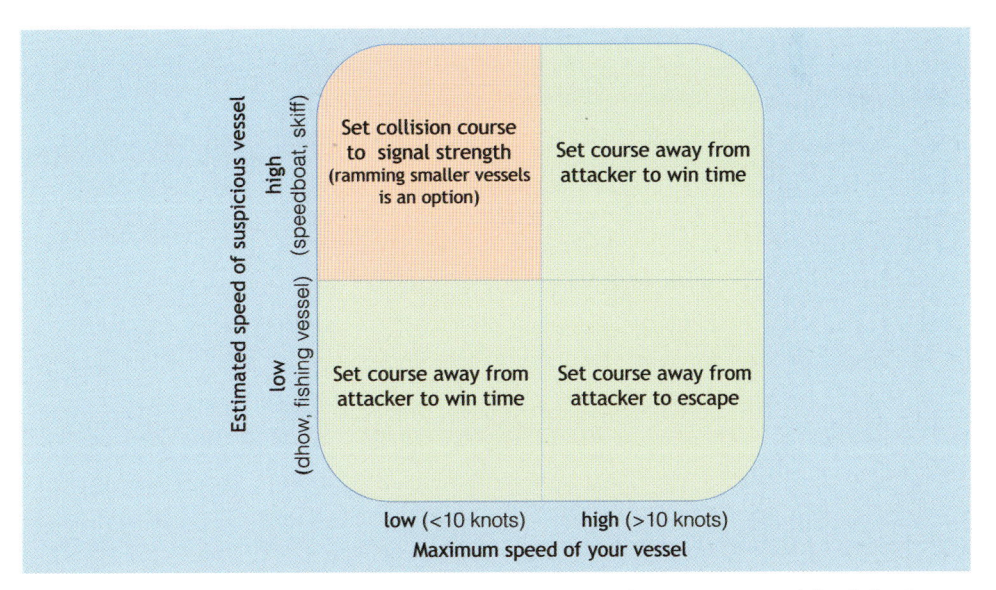

	low (<10 knots)	high (>10 knots)
high (speedboat, skiff)	Set collision course to signal strength (ramming smaller vessels is an option)	Set course away from attacker to win time
low (dhow, fishing vessel)	Set course away from attacker to win time	Set course away from attacker to escape

Estimated speed of suspicious vessel (vertical axis) / Maximum speed of your vessel (horizontal axis)

FIG 4.20: Course options dependent on the relative speeds of the attacker and the defending vessel.

To summarise: all of the crew should be alert and prepared to initiate countermeasures whenever any boat is approaching fast in a straight line (over any distance) or is close to or within the perimeter, and whenever you plan to get close to a coastline.

A suspicious vessel is any vessel that is not clearly harmless. The latter are any official ships showing national insignia, military ships and large commercial vessels. Whenever another vessel is approaching, consider it an offensive approach and raise the alarm.

Be mindful of the fact that an offensive approach does not necessarily need to be an attack. You cannot be sure of an attack before the boat reaches your 50-metre perimeter (despite your warnings and actions) or you identify the preparation of weapons or boarding tools.

Whenever the alarm is raised, all crew should muster on deck. The only exception may be the crew member on communication duty below. Kids and all crew incapable of carrying out any duties during the following stages should retreat into rooms below the waterline. With the crew on deck you demonstrate that the approaching boat has lost the element of surprise and that the possible target is ready to react and defend itself.

All equipment for deterrence and defence should be readied (if it is not already at hand).

Start your engines and consider using them to support your yacht on the course chosen for preparing (or evading) a confrontation. Should your sails not lead to increased speed on your chosen course, drop them completely or partially (headsail taken in, mainsail close-hauled) so your yacht is as manoeuvrable as possible.

The two major alternative courses for a yacht are determined by the weather, the sea state and the relative speed of the approaching boat and your yacht.

The best option is to evade the attackers and avoid even the chance of a confrontation. Consequently, if you estimate your yacht's speed is capable of a similar or higher speed than that of the approaching boat, plot a course away from the attacker and try to put as much distance between you and them so that you can eventually escape.

Many opportunistic and inexperienced pirates use large wooden boats or smallish tubs with relatively weak engines. An escape from these boats is feasible if the manoeuvre is initiated as early as possible. At least one sailboat is recorded as having outrun a suspicious local wooden motorboat with a speed of 10–14 knots in Latin American waters.

Some skippers have also reported plotting a course that presented their beam to a rather high wave. Although uncomfortable, the seas were easily taken by their sailing yacht and the apparent wind directions were nearly optimal for their speed. In contrast, lacking a keel, the approaching open skiff could not bear those waves very well. The rolling motion was said to have handicapped all activities on board, making a boarding from that ship nearly impossible. Even if you do not manage to escape the vessel, you will win precious time. Always check whether averting your course or escaping is an option.

In the event that the pirates approach with heavily motorised skiffs, a yacht will most likely not be able to escape. Some boats used by pirates are capable of speeds of up to 25 knots. Of course, you could still try to run from such boats to win some time, but an alternative may be a show (or at least the pretence) of force by turning against the approaching vessel. Point your bow directly at them and demonstrate that you will be offensive. If your yacht is robust and their boat is weak, you may consider ramming the approaching vessel if the crew is identified as armed and hostile.

However, even a 40ft sailing yacht with a determined crew on deck presenting (improvised) weapons or replicas will have an effect on a crew in a skiff. Not many pirates in an open boat like the thought of being shot at or rammed by a larger and far more robust boat. Boarding a boat passing you at high speeds is more or less impossible, too (although the skiff may turn and then attempt to approach from the aft sector).

That said, please be advised that this tactic is a last resort tactic, preferably of a well-armed crew against **opportunistic pirates** only. It would be very bold and dangerous for any yacht to confront a fully manned and heavily armed vessel of professional pirates. They may well shoot your boat to pieces before you get close.

If your AIS transmitter was turned off, activate it now.

One crew member should contact the approaching vessel on VHF channel 16 to demand it immediately aborts its approach. Threaten to use weapons against them, even if you have none on board. Pretend to be stronger than you are. Try to stay relaxed when speaking. Don't sound afraid. Instead, demonstrate a professional cool. It is very helpful to have practised a good text, such as the one in the box here, a couple of times before you embark.

After sending this message two, not more than three times, the respective crew member should join the team on deck.

Depending on the speed of the approaching vessel, it may already be

> **VHF COMMUNICATIONS: PRETEND TO BE STRONG**
> *'Unknown vessel, unknown vessel. You are approaching the armed vessel* Stingray. *Abort your approach immediately or our security team will engage you.*
> *I repeat: abort your approach immediately or you will be shot at. Go away! Now!'*

closing in on your 500-metre (-yard) perimeter and you will be able to identify first details without binoculars. This is, of course, also true for the other crew. Professional pirates have been reported to start shooting from a distance of about 350 metres (yards) to intimidate the target vessel and make it stop. A distance of 500 metres (yards) is therefore when you should start your deterrent measures. Since the approaching crew should be able to see them, this may result in them aborting before they close in any more.

500 metres (yards): deterring an offensive vessel

If the approaching vessel has not yet shown unmistakable hostile actions (preparation or use of weapons, preparation of boarding tools), you must try to avert their course by means of deterrence to avoid a confrontation.

As stated above, it will be helpful to show as many crew on deck as possible. This signals that the attackers have lost the element of surprise and will face possibly stronger resistance than anticipated. Your team will also be more able to deter and defend when mustered on deck.

If you sail a region that is known for the use of security teams on board, a simple trick can be very effective (actually, it is likely to work in other regions, too): dress up and pose like a pro. Although you will most likely not have a battle helmet and Kevlar vest on board, you can easily pretend to be a security guard. Swap your bathing trunks for cargo pants, Hawaii-patterns for camouflage. Drop your colourful surf-brand shirt and don a black T-shirt instead. Exchange your sailing cap for a military-style 'soft cover hat'. Dig out those sailing gloves (or the working gloves you use while handling mooring lines) and put them on when the bad boys approach.

You will be surprised how the impression of your crew will change when viewed from another vessel. You can buy this kind of outfit at any surplus store or market very cheaply.

All weapons and replica should now be ready for presentation or action.

Your emergency signals should be ready at hand to be used by the crew on deck within seconds. These include your EPIRB, the DSC-VHF radio, PLBs, parachute flares, the satellite phone with coast guard or police on speed dial, and a DSC-HF radio. You may not be able to operate all of these from deck, but try to have most of them ready. On a high-risk passage, it would be advisable to have the EPIRB, PLB, parachute flares and a fully charged satellite phone stored on deck at all times.

Only if your crew numbers more than five (not counting young children and the very

FIG 4.21: *Disguised as an armed guard in the Gulf of Aden, this sailor is wearing military-style garments from a Chinese store and holding a self-made replica rifle. Two suspiciously approaching boats turned away at 700 metres. (Photo: Berswordt)*

elderly) should one member constantly operate the communication systems below deck. Remember: you want to show strength on deck.

Always keep your eyes on the approaching vessel. The moment you identify hostile activities, you should initiate defensive measures (see pages 81 and 91–6). All emergency signals will be deployed in this case.

If you are still deterring, act in a determined, unmistakable and offensive manner. Don't be shy! The more resolute you are, the more likely it is that any opportunistic attacker will abort their approach.

In this first stage of deterrence, you are using visual signals. The most effective measure would be the presentation of weapons or well-made replicas. The more powerful, the better (see 'Part III' for ideas and inspiration).

To show their determination, professional security teams take a position at the guardrail and hold any weapons either up over their heads or clearly visible at the side of their bodies. It's important that the weapon's silhouette can easily be identified from a distance. Try to have the sky or an even surface behind you; cluttered backgrounds may make the weapon hard to recognise. Just standing there alone may do the trick.

If you have more crew than weapons and replicas on board, the others should make unmistakable hand signals and acoustic warnings. Officially, a sequence of continuous 'short-long-short-long' sounds from your fog horn means 'Stay Away'. Not all skippers from remote areas of the world will know the signal, but they will notice that the crew is alarmed and that they are trying to convey some kind of message. Issued together with hand signals and visible weapons, the intention should be quite clear.

There are several reports of crews showing such determination who were able to effectively deter offensively approaching boats, not manned by professional pirates. We are not aware of a single one that did not succeed.

200 metres (yards): offensive deterrence

If a boat approaches past the 200-metre (-yard) mark after noticing your efforts at 'soft deterrence,' you need to put some more punch into your measures and be ready for a confrontation. In some regions, professional pirates wielding rifles and assault rifles are known to have opened fire to intimidate the crews of target vessels shortly before reaching the 200-metre (-yard) mark.

Check one last time if your course is favourable for a confrontation and compensate, if necessary (we assume that you have a working autopilot installed).

At this point, call a MAYDAY with all available signals. Deploy your EPIRB, push the distress button on every communication system that has one and try to contact the local coast guard if you have their number and if you can spare a crew member.

Start to shoot a parachute flare or flare guns (vertically at this point) regardless of whether it is day or night-time. The purpose of the signals is also to show the approaching vessels that you are worried and ready. It will serve to unsettle the attackers as they have to assume that others have noticed these and other signals from your vessel. Depending on the remoteness of the area and the general presence of official units, they will have to think about ships or aircraft approaching at the scene.

If you have them ready, you should deploy prop-fouling devices after you have made your last intended change of course.

Should you have weapons on board, you can now also start firing warning shots at the attackers. Do not merely shoot into the air. Aim in front of or over the heads of the approaching crew. Your aim should be true to make sure that you do not hit the boat and injure any crew before they show clear signs of hostility.

If you have no 'real' weapons on board, you can improvise them (see Part III for details). Shots from flare guns and parachute flares shot horizontally make impressive warnings for approaching wooden and plastic boats. Be aware that many models of parachute flares expel parts and all expel hot gas from their rear end. Be certain that the area behind anyone firing such a rocket is clear and never hold the device in front of your face to aim. Having to aim them from a position next to your head makes it hard to hit a target over longer distances.

50 metres (yards): deliberate defence

If the approaching vessel does not alter its course despite the deterrence measures you have taken, you have to be ready to repel a boarding, even if you cannot see weapons or boarding tools being readied at this time.

All defenders should take their next actions from a position of cover if your boat offers it (see Part III for details of safe positions on your boat). Try to at least conceal parts of your body from the attackers to reduce the area that can be aimed at.

Before going into details of any potentially lethal action, we need to remember that many nations do not allow disproportionate violence against attackers. Let us look at a land-based example: you face serious legal problems in Germany if you shoot and kill a burglar armed with a knife within your house. The law would expect you to only disable the offender. As the national law of the yacht's flag state is valid outside any 12-mile zone, it is up to you to find out about the laws of self-defence in your flag's country.

We strongly recommend that you are prudent when using any weapons against fellow humans. Whether it be in court or when you look at yourself in the mirror later, proportionate defence will be a positive effect after the event.

Consequently, if the vessel carries on approaching you despite your warning shots but without showing weapons or boarding tools, you should consider deliberately aiming your weapons at their rudders and propulsion system (engine, drivetrain and propeller).

If you are already under armed attack, using violence to defend yourself will not be an issue in most countries: if someone shoots at you, you are allowed to shoot back. This situation would be a clear piracy assault and your methods to fend it off would be deemed self-defence. Most countries allow lethal violence against other crews only in the event that they use lethal weapons themselves or if allowing a boarding would lead to the risk of severe injuries or death.

The same holds true if you use improvised weapons, such as your flare gun, parachute rockets, magnesium flares thrown by hand, machetes or whatever you have.

To date, there is no precedence of a crew who successfully and maybe violently fended off armed pirates being prosecuted by any court.

Fending off boarding parties

The time of boarding is most likely the last opportunity to actively avoid robbery, kidnapping or murder. Again, well-prepared, alert crews on better-equipped yachts will have more chances to succeed at this last line of defence than a crew who was taken by surprise and has to resort to last-ditch measures.

Maybe the crew has decided against any defence at the boat's sides. They should now raise their empty hands and give themselves up to the attackers and an uncertain future. However, in areas plagued by professional pirates those crews should use any time remaining to prepare the boat for a subsequent rescue by military or police (see page 160).

Crews who opt to fight are now taking the highest risks of the confrontation. From a close distance, even pirates wielding 'only' revolvers or pistols can bring them into accurate and lethal action. Some of the attacking crew will most likely try to threaten and control your yacht from their vessel. Another group will try to board.

Consequently, the defenders will have to cope with two groups. This is a major challenge for an untrained, let alone an unprepared, crew. On the one hand, you will have to dodge or fight the pirates in their vessel, on the other you have to fend off any criminals trying to climb aboard your vessel.

If your crew is large enough (actually more than one), you may want to counter both threats at the same time with one part of your group attacking the boarders and the other aiming at their vessel.

Apart from firearms, the best measures employed against the vessel will be flares and parachute rockets. All of these incendiary devices pose a huge danger to any boat not constructed from aluminium or steel. If you score a hit on a wooden or GRP boat, you can rest assured that the pirates will have other troubles than harassing you. Remember, though, that the fire could spread to your boat, too.

The harder it is for the attackers to climb on to your boat, the longer they will be exposed to your crew's defensive measures. If they only have to take a big step from their skiff to you're a swim platform while aiming their weapons at your crew, they will be in control in next to no time. The situation shifts rapidly in favour of the defenders if the attackers have to overcome improvised barriers or even cut barbed wire or electrified fences. Their attention will be focused on the barriers and they may need both their hands to deal with the obstacles. For precious seconds or even minutes, they will be very vulnerable to your defensive actions.

Your crew will be able to counter their efforts with firearms, thrown missiles or melee weapons. Obviously, although firearms are the most effective weapons in such situations, they are not the only tools: pepper spray from large cans with long-range jets are capable of effectively disabling even armed pirates.

One report tells of a crew fending off pirates in Latin America with long wooden boathooks and wooden poles. As they were prepared to fight back and better armed for the situation at the boat's side, they were successful against four attackers armed with machetes.

What to do after being boarded

The phase after being boarded and controlled is usually the hardest to be endured by victims of robbery or kidnapping. You will be at the mercy of a group of hard-to-predict, possibly drugged, criminals. Depending on your captors' motives and capabilities, this phase will last just a couple of minutes (the time necessary to strip you of all your valuables) or many weeks and months.

However, both police and military units that specialise in hostage situations at sea recommend you cease all hostilities as soon as you realise that a boarding by pirates armed with firearms cannot be avoided. In areas in which a kidnapping is probable, you should use any time remaining before being overcome to prepare for a rescue by professionals.

Crews who are trained in military tactics using firearms may be the exception and could try to fight armed pirates after they boarded. Of course, it is generally possible that a civilian crew can score a lucky punch against pirates in control, but fighting armed criminals after losing all initiative will be a huge risk and any defeat is usually very harmful for your crew.

In line with military professionals, we recommend that crews stop their resistance once pirates armed with guns are securely on board.

Rescue by military and police units

Preparing for a military intervention

In some areas of the world the persistent threat of piracy has been met with the deployment of larger units of military or coast guard personnel. Typical examples in recent years were the northern Indian Ocean (international naval forces) as well as the Strait of Malacca (coast guard units from Singapore, Malaysia and Indonesia). Pirates who operated in these areas also had a high tendency to kidnap their victims for ransom. Some pirate will try to claim your vessel, others will transfer the crew to theirs.

> **WHEN YOU CAN'T FEND OFF THE BOARDERS**
> - *Render ship unable to manoeuvre and move (block rudders, disable sails and engine)*
> - *Block all main entrances in an open position to allow for easy access by the rescuers*
> - *Block the transmit button of your VHF-radio in active position (hot mike)*

To handicap pirates choosing to remain on your vessel, it is useful to deny the attackers valuable capabilities of your yacht and prepare it for boarding.

European forces specialising in anti-terror and anti-piracy recommend actions that aim at supporting your yacht's detectability, hinder its navigation and weaken the pirates' defensive positions. A crew familiar with the steps on a prepared vessel can execute the necessary actions in quite a short time.

By the time pirates board your vessel, you should have already deployed the EPIRB – maybe you have even dropped it on a line at the bow (making it harder for boarders to notice the device, since they usually attack from the stern).

Discard all scruples you have developed during those SRC classes. Block your radio's transmit button with a piece of tape on channel 16. This improvised 'hot mike' switch will transmit all conversation on board to anyone listening and it produces a signal that can easily be used by the authorities for fixing your position.

Disable the rudder. Consider throwing your steering wheel overboard (pirates may not know that you have an emergency tiller on board) or cut the wires of your steering.

Disable your vessel's means of propulsion by cutting the wires to the starter batteries and cutting the most important halyards as close to the mast as possible.

If the weather permits, block the doors to your saloon (catamaran) or the companionway in an open position. This means your attackers will have a hard time barricading below decks, and gives your rescuers easy access to all parts of the boat.

What to do during rescue operations

Any military rescue operation is a very dynamic and very dangerous process for all involved, including the yacht's crew – whether in hiding or being held hostage, they are under threat both from possible retaliation by their kidnappers as well as from stray bullets fired by the rescuing team.

In at least one case, the capturing of the US-flagged sailing yacht *Quest* in 2011, the pirates shot and killed all hostages when US warships appeared on the scene. In another case in 2009, a crew member aboard the French sailing yacht *Tanit* was accidentally shot by the military during the rescue operation.

There is no real cover from bullets fired by modern assault rifles on a GRP or aluminium yacht: all decks and walls will be easily penetrated and the area of operation is extremely small and crowded.

Consequently, the best position to be in is lying down, either flat on your belly or in a foetal position to reduce the area that can be hit. It will be very helpful if you could present your hands visibly to reduce misunderstandings on behalf of the rescuers. Remember that the good guys have only a split second to decide who is a criminal and who is a hostage. Any mistake on their part may end fatally for one of your crew.

Anti-terrorist specialists are adamant that you do not attempt to support them during their actions. The risk of injury, either from them or the pirates, is far too great and usually their tactics do not rely on any active support from the hostages.

The sequence and equipment (e.g. RIBs, helicopters) used are mission specific and hard to predict. In the event that a sailing yacht is involved, a helicopter may be deployed to spot you and harass any pirates. The rescuers will most likely board your yacht from a boat, as it is dangerous to fast-rope from a helicopter to a sailing vessel with an intact mast. If your vessel is in motion, the rescuers may use warning shots to convince the pirates to stop. It would be wise and helpful if all crew hit the floor at that time (if pirates allow it or are past caring).

The crew should stay in the prone position until they are asked to get up by the rescuers. To get up earlier could lead to a fatal misunderstanding. In some cases, rescuers might even handcuff the hostages and separate the goodies from the baddies once they are in full control of the situation. Do not resist or act up and try to keep in mind that the situation is extremely dangerous for the guys risking their lives for you, too.

FIG 4.22: *Indian Navy helicopter checks out a boat in the Gulf of Aden during a regular airborne patrol. (Photo: Berswordt)*

SECURITY IN ACTION

The situation is not safe until the rescuers have positively identified the crew and reliably controlled all of the pirates. The rescuers will tell your crew when it is safe to get up and move about.

SAILING IN A CONVOY

Successful Convoys at a Glance

A well-coordinated convoy will considerably improve your security in many regions. Although it may be unwise to rely on sailing in a group to protect yourself against determined professional pirates and terrorists, a well-sailed convoy will deter most other criminals. There has not been a single attack on a tight-sailing convoy reported to date (this includes regions known for pirate activity). Consequently, whenever you are contemplating a passage in areas that are known for offshore criminals, sailing in a convoy may be a good strategy.

> **CRITICAL SUCCESS FACTORS FOR CONVOY SAILING**
> - *All skippers agree upon goals, priorities and the region's risk assessment*
> - *The participating boats and crews are of similar performance and capabilities*
> - *Boats are sailing at a distance of no more than 1.5NM apart*
> - *During the passage, crews stick to the commitments made during the preparation*

Pros and cons of sailing in a convoy

Sailing in a convoy comes with some pros and cons that you may want to consider when planning your individual cruise through high-risk areas.

The mere presence of a group of yachts will deter some types of attacker. An intensified lookout will increase the probability of an early detection. Convoy ships could also support each other when one of them is offensively approached by means of offensive manoeuvres or at least communication with coast guard units while the target ship focuses on dealing with the approaching boat. In the event of a successful boarding of one convoy member, the remaining boats can offer medical and psychological support, shadow the kidnapped boat from a safe distance and coordinate rescue missions with the authorities.

Irrespective of the obvious security benefits, some other reasons may lead a group of skippers to sail together. They will support each other if there are technical problems or low provisions; weather forecasts can be obtained only once for the group and routing options can be discussed; and reciprocal visits on the yachts and daily radio communication are fun,

> **PROS AND CONS OF SAILING IN A CONVOY**
> ✔ *Strong numbers deter attackers*
> ✔ *Improved lookout*
> ✔ *Support during approach, boarding and after the attack*
> ✔ *Technical and logistic support during the passage*
> ✔ *Fun while sailing together*
>
> ✘ *Additional time invested for coordination*
> ✘ *More active manoeuvring to keep convoy formation*
> ✘ *Generally slower progress*
> ✘ *Necessary radio communication between ships (increased risk of being detected)*

positively distracting and generally lead to an increased sense of safety and comfort without compromising the actual security.

However, successfully sailing in a convoy requires additional communication before setting sail and more coordination while under way. To keep an effective convoy formation, all ships involved need to execute more manoeuvres than they would if they were sailing solo. Especially small crews may be reluctant to plan for added work to keep the right position in a convoy. A slightly increased risk of being detected results from the necessary radio communication between ships (unless you intend to signal with flags and Morse code light signals).

Overall, though, a convoy is usually more advantageous than sailing solo if you succeed in compiling a group of harmonising crews on sailing boats of similar performance.

> **SAIL A SUCCESSFUL CONVOY, AGREE ON THE BASICS**
> • *Share individual risk assessments*
> • *Define ports at destinations and stopovers*
> • *Minimal level of mutual support to be offered should there be an attack*
> • *Departure priority (date or weather)*
> • *Course priority (planned course or weather)*
> • *Maximum and minimum speed, use of your engine*
> • *Convoy formation, including maximum and minimum vessel distance*
> • *Condition for separation and break-up*

Preparing the convoy

While preparing a convoy and sailing in a group is not rocket science, you should take your time to conduct some simple pre-checks and ensure that everybody in the convoy has the same goals and expectations. Overall, most successful convoys are compiled and coordinated in less than two days. In convoys of pleasure boats with civilian crews, it is important that the skippers feel mutual sympathy, trust and respect. Therefore, it is helpful if you spend sufficient time together to find out whether or not the group harmonises.

Investing in an extra day of sailing to practise formation sailing and convoy manoeuvres before actually heading out was reported as being helpful by some crews, whereas many others simply sailed some manoeuvres in the first hours of the actual passage to save time.

First things first: let us take a look at the necessary preparations.

Goals of the convoy and skippers' expectations

The first step of a successful convoy-planning phase is the understanding and matching of the participating skippers' motivations and goals (bearing in mind that skippers represent their crews).

Many convoys sail ineffectively or are broken up before reaching their intended port because differing requirements and motivations are not identified and addressed before setting sail. As a result, a convoy should not depart before all relevant matters are on the table and their implications for the journey are understood and accepted by all.

Agree upon an open and respectful discussion. Keep in mind that people cope with real or perceived risks in very different manners. Many skippers and crew will be quite emotional about discussing matters of security in such demanding environments. Aim to keep the discussion at a factual level while respecting the fears and insecurities of all participants. A civilian convoy passage will only be a success if all participants understand and respect the views of all others.

For a good start, each skipper should share his or her view on the perceived risk in the area to be crossed. Do they expect opportunistic, inexperienced attackers or professional pirates, let alone terrorists?

Participants understanding the expectations of others understand their priorities much more easily. A skipper anticipating a probable terrorist attack will be more willing to start his engine for an optimised speed than someone who does not expect more than aggressive fishermen. Recognising the perceived threat situation vastly simplifies the discussion of all following topics. Different estimates are not a problem per se as long as all members are willing to bear the consequences together (e.g. higher fuel consumption, increased time for the passage, etc).

After talking over risk assessments, you need to discuss the actual port of destination for every ship and any planned stops to be made along the way. Even in apparently risky areas, stopovers could be made to visit attractive locations. More often, they will be necessary logistical stops for some yachts to bunker fresh water, fuel or provisions.

The next topic should cover the convoy tactics and mutual support to be used should a potentially threatening vessel be detected or offensively approach. There is a simple tactic to eliminate the occasionally overwhelming emotions that can arise while discussing this issue: rather than discussing the right tactics on a very abstract level, it will help if every skipper shares her or his personal ideas and what they deem to be the limits when it comes to supporting others:

- Would you be willing to make the effort to close ranks the moment a vessel is detected?
- Would you set a course towards an attacked convoy ship to demonstrate support the moment an offensive approach is identified?
- Would you help in calling for help (do you actually have long-range gear on board) and as dispatch if there is a confrontation?
- Would you even plot and intercept the course, ramming an attacker to prevent the crew from boarding one of your convoy mates?

After sharing their views, the group should agree upon the minimal level of support and discuss whether the upside of such added safety actually justifies the increased coordination required to travel in convoy. If the convoy participants are not even willing to close ranks upon the detection of a suspicious vessel or change their courses in support of an approached convoy mate, sailing in a group may be a bad idea after all. The only upside for the convoy is the hope that potential attackers are deterred by the mere existence of other boats on the water.

Your group might now continue by sharing their priorities concerning the date of departure. Some skippers may be forced to leave on or before a certain date (e.g. because visas are expiring), while others might prefer to wait for perfect weather to reduce fuel consumption. Choosing a date of departure that is not accepted by all skippers could lead to a lot of dissonance during the passage as it may result in continuously rising costs (e.g. fuel consumption) or reduced comfort (e.g. head winds, gales, waves). These issues that affect the ships over a longer time during the voyage have been reported to cause equally long periods of frustration. If everyone has agreed upon the same date of departure with all consequences in mind, such continuously nagging effects are not to be expected.

Similar emotional mechanics are at work for your next topic, which we shall label as 'course priority'. Depending on their risk assessment as well as the performance of their boat, some skippers may try to stay on a course plotted prior to the departure. When the weather turns against them, this group may prefer to start the engine to remain on course. Others may wish to be flexible in order to optimise the passage in relation to the development of the weather, even if it leads them closer to a dangerous coast. Perhaps it is their goal to maximise speed or minimise fuel consumption. Whenever the passage that your group plans will take longer that the period covered by reliable weather forecasts, you should generally agree whether your group will stick to the original course or flexibly react to the weather.

The same holds true for the group's 'speed priority'. During the discussion of this topic, the general performance data as well as the capabilities and priorities of the crews should be shared openly. Having these characteristics at hand, the group should agree upon a maximum as well as a minimum speed for the passage. Crews of the higher-performance boats might need to reduce their speed (not really an easy thing to accept if you expect an assault or just hope to save a day of provisions). Crews of the slower-sailing vessels risk having to start their engines more often to keep up with the convoy. The decisions regarding the upper and lower speed limit consequently affect both the provisioning and bunkering tactics of each boat. Obviously, it is possible to change the details of your strategy when sailing. Nevertheless, it will be useful to know your priorities and tactics as a basis for your provisions and fuel plans. Should you disregard this topic in favour of discovering and discussing any differences in speed priorities and abilities while en route, your convoy is at strong risk of breaking up before reaching its port of destination.

Before discussing the concluding topics, skippers should share details of their relevant pieces of equipment to establish a basis for later support. Such equipment could be the type and volume of fuel, fresh water capacity and desalination abilities, brands of major equipment and spares carried, communications systems and emergency equipment, weapons and replicas. You may also want to share details of your map provider, source of weather forecasts, navigation software and navigation system brands. Knowing differences in systems and data sources helps to keep discussions on routing efficient while sailing.

Finally, it will be a good idea to briefly discuss the factors that could lead to the convoy's break-up. Sometimes skippers misjudge the capabilities of their vessel or crew during the preparation of the convoy. After departure, their boat might start to show some deficiencies and become the convoy's 'drift anchor'. Combine such developments with differing views on the general risks in the area and you will have the ingredients for emotionally charged conflicts. By contrast, if you have discussed the circumstances that may lead to the separation of single boats or the break-up of the convoy, you will depart in friendship and avoid negative vibes throughout.

Discussing these topics also helps to uncover any fundamental issues, such as some skipper's opinion that he or she would not have sailed without being part of a convoy in the first place.

Compiling your convoy

After completing the discussions for preparation, your group will have a clear picture concerning boats, crews, goals and performance. You will also have a good idea of whether or not you harmonise well enough to rely on each member of the convoy.

Convoys compiled from boats with similar goals, priorities and performance will probably be successful formations. The same holds true for groups who decided to equalise differences in performances by regarding the weaknesses of some boats. In both cases, skippers need to mutually like and respect each other. A good relationship between them is quite essential when the group needs to flexibly react to changing conditions. Ones who like each other will also be more likely to support their counterpart in emergencies. If your group perceives differences that are too major to be compensated for, you may consider sailing separately.

Convoys comprising two to four boats are ideal in that they maximise the security effect while minimising efforts for communication and manoeuvres during the passage. A convoy of three ships is ideal for skippers who are sailing their first convoy. It may be wise to divide a larger group into two or more separate convoys. If you are able to match boats according to their goals and performance, the smaller homogenous convoys will be more successful than one large heterogenous group.

Groups comprising ships with different performances (i.e. speed while motoring as well as speed and course while under sails) will be a problem if the skipper commanding the stronger boat does not want adapt to the speed of the smaller, lower-performing boats. By contrast, when skippers agree to mutually regard the other boats' weaknesses, sailing together will not be an issue. It is essential that the agreements are respected throughout the passage. This will be demanding during lengthy periods of slow sailing (for the skippers of fast boats) or motoring (for the skippers of the lower-performing vessels), but skippers should try to silently stick to the commitments that they have made. The general convoy strategy should only be altered if external factors were to change, such as weather or some relevant news from the outside about a new security situation along the route. On any other occasion, discussing the strategy while en route places stress on a good convoy atmosphere.

Convoy routine: communication and coordination

The rather small distances between convoy sailors make a standard VHF radio the system of choice for convoy communications. Generally, you should try to limit any radio communication to avoid revealing your presence in the area.

Before setting sail, skippers should agree upon the frequencies and time slots for convoy communication. It will be most convenient to communicate twice a day to share speed and course preferences for the following 12 hours. Obviously, any other communication – such as discussing the weather forecast, information on security, technical issues or simply small talk – can take place during these conversations.

Many skippers are happy with one time slot after dawn (for your daytime tactics) and one before dusk (to prepare for the night). The exact time will be dependent on the position of the convoy as well as the crews' preferences. When agreeing on a time, have your watch rhythms and general on-board routine in mind; you do not want to have this disturbed too much by the radio calls.

So long as you keep as close together as good convoy tactics demands, you should consider setting your radio to low transmit power to reduce the propagation of your signals and limit the radius in which you give away your position. Bear in mind, though, that even low-power signals will carry 5NM or further. Although such a distance is within visual range in clear weather, it would be safer to rely on hand-held VHF devices set to low power, if available.

Although it is possible to choose a network coordinator for such communications, most groups do not need one: everybody just needs to call in at the agreed time and share their views and information.

Obviously, the language used for such communication will be determined by the preference and capabilities of the participants. It may add a little extra security if you exchange your information in a language that is rarely spoken in the area in which you are sailing, since this will reduce the chance of a local criminal understanding your course and speed plans.

Under no circumstances should position reports be shared via un-encrypted VHF communication. No eavesdropper should ever obtain the convoy's position. If the convoy vessels lose visual contact but are still able to talk on the VHF, they should employ simple encryption methods. One idea would be a pre-defined multiplier that is applied to the coordinates (or parts thereof). Another system involves the use of a fixed reference position only known to the convoy participants. Rather than sharing their latitude and longitude, crews share just their distance and bearing in relation this (secret) reference position. Most criminals are unlikely to possess the technical and intellectual capabilities to deduce your true position from this publicly shared position.

If all ships can send and receive e-mail via satellite or PACTOR, sharing positions in mails is a safe option for the convoy.

For convoys equipped with it, using HF/SSB rather than VHF would be another alternative. You would use low frequencies with the lowest-possible transmit power. Since there are so many frequencies, it will be difficult for an accidental eavesdropper to detect your communication. Bearing in mind that HF carries quite far, the potential area they'd have to search in order to find your true position after receiving a signal is also extremely large. Unless you have a technically very capable criminal (or you give away your true position), detection is very unlikely.

AIS is a very tempting technology to help to constantly coordinate your convoy. All ships can see the other's positions, speed and course at all times. On the other hand, the same holds true for any offender in a range of 25NM or even a lot farther, since AIS signals are known to carry for 100NM and more under some atmospheric conditions. The skippers will have to weigh the pros and cons and make a decision. In several regions that are under tight surveillance by local authorities (Strait of Malacca and the Horn of Africa, at the time of writing), AIS is a good option. High-risk regions without such security cover (e.g. the coasts of Venezuela, Guatemala and Honduras) are better sailed without it so that yachts remain as stealthy as possible.

Convoy routine: lookout

As described in the 'Detecting the threat: the effective lookout' section (see pages 150–2), a keen lookout is the essential basis of all of anti-piracy tactics.

A convoy offers some options to reduce the workload that an effective lookout entails for individual crews. If your boats are sailing in a tight formation, each boat could take responsibility for a time slot agreed in advance. This would allow the other boats' crews to relax a little during that time. As an example, each boat in a three-vessel convoy would be on lookout duty for three hours before passing the responsibility to the next boat, thereby enjoying six hours of reduced vigilance. Be aware that such a shared lookout only works in tightly sailing groups.

Another tactic used by convoys allocates a discrete sector for each boat according to its position in the convoy. The lookout is eased for each crew because they do not have to continuously watch and assess an area spanning 360-degrees, but rather a fraction of that.

If everybody reliably watches their sector, no vessel will be able to close in on the convoy undetected. The formations shown here are described in the 'Convoy routine: formations' section (see page 171).

To be extra safe, you may want to forget about all of these tactics and have every boat watch a 360-degree sector all of the time. This might be a good tactic if the crews on boats are generally large (> four able watch-keepers).

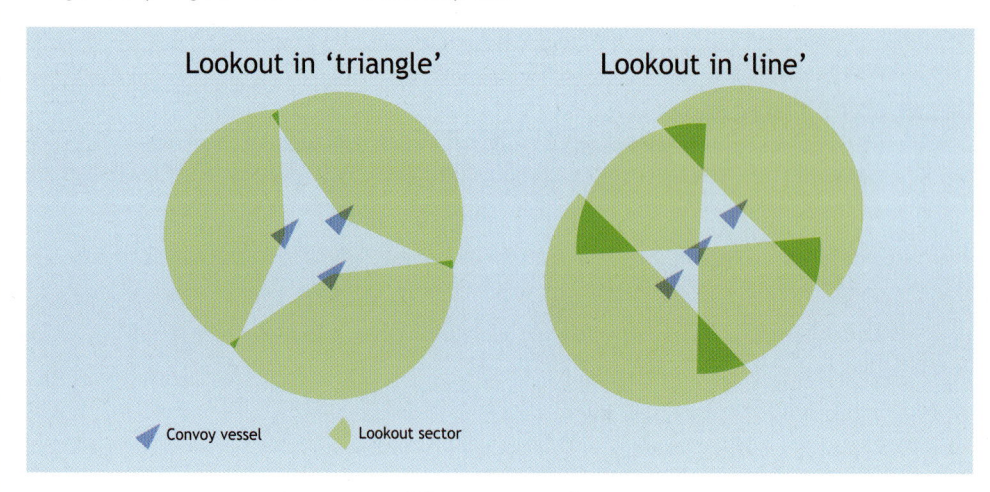

FIG 4.23: *Distinctive watch sectors in different convoy formations.*

Convoy routine: distance between boats

When considering the distance that the boats of your convoy should keep during routine cruising, two factors are important: collision avoidance (internal security) and deterrence maximisation (external security).

To maximise your effect against potential attackers, a tight formation is best since it means assailants have to be ready to cope with more than one vessel at the same time, and this is off-putting. This effect is lost whenever the attacker does not perceive the boats as sailing together.

To be regarded as a single entity by a potential attacker, boats should sail no more than 1.5NM apart. Whenever your boats spread further apart, the group does not really look like a convoy any more and this may encourage attackers to attempt an assault on a single boat. Another disadvantage is that being far apart means that it would take longer to reach one of the boats and provide support if there were an assault.

You should use the boats' top speed with running engines as a basis for your estimations concerning spacing of the convoy. To allow for effective support after detecting a suspicious vessel, no ship should take longer than ten minutes to get alongside any other convoy boat at any time. Consequently, if the top speed of your slowest boat is 7 knots, the maximum distance should be 1.2NM.

The minimum distance is only limited by the crews' capabilities. Back in the 18th and 19th centuries when convoys of sailing vessels roamed the seven seas, a distance of one cable length (185.2m or 1/10NM) was the desired distance between ships for maximal mutual support. Obviously, these were manned by large numbers of mostly professional sailors and thus capable of constantly monitoring and adjusting the course and speed under sail in order to avoid collision. Small crews on modern yachts will face difficulties safely sustaining such a short distance, as continuous changes in course and speed will be very tiresome. A distance of 0.4–0.75NM seems to be manageable in fair- to moderate-weather conditions. If the weather or sea state deteriorates, crews should consider increasing the distance.

Convoy routine: formations

During a routine convoy operation, two formations have proven effective and rather simple to maintain: the 'triangle' and the 'line'. The 'front' formation would only be chosen as an offensive measure if a suspicious vessel deliberately enters the safety perimeter.

Many boats sailing in convoys consider the 'triangle' to be optimal. The distance between all boats is about the same, allowing each to assist all others equally quickly. There are no boats directly before or behind any other vessel, which means that other boats do not have to react to each speed change of the preceding or following ship to avoid a collision. Accordingly, this formation allows the group to 'breathe' within the limits of the maximum and minimum agreed distances.

172 THE COMPLETE YACHT SECURITY HANDBOOK

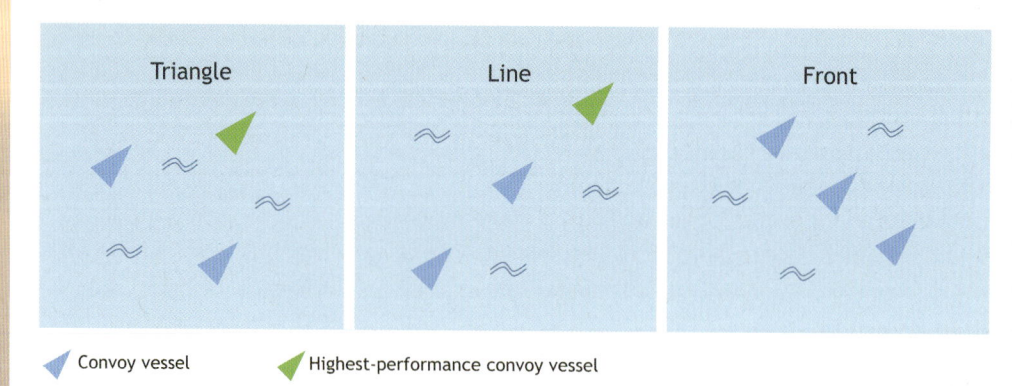

| Triangle | Line | Front |

◄ Convoy vessel   ◄ Highest-performance convoy vessel

FIG 4.24: *The two most useful formations for routine sailing: the 'triangle' and the 'line'. The 'front' formation is useful as an offensive pattern.*

Generally, the better-performing boats are responsible for keeping the distances correct. Whenever they notice that slower boats cannot keep up, they should reduce their speed accordingly. The best-performing boat will also usually take the lead in this formation. It is the leader's responsibility to ensure that he does not race too far ahead. Maintaining a strict formation is not necessary all of the time, though. The leader may reduce speed to actually invert the triangle or sail in a front formation.

If you are building a convoy of four boats, the triangle may be expanded into a diamond shape, which offers most of the triangle's advantages over other formations.

The 'line' formation may be a good alternative when you have to cruise in tight passages, such as those found in between islands, within straits or while travelling in a traffic-separation zone. Since straits and archipelagos are pirates' favourite spots for ambushes, using this formation allows all convoy ships to keep a maximum distance from the dangerous coastlines. The line's disadvantage is that it is inflexible: whenever one boat changes its speed or direction, most others have to adjust swiftly to avoid collisions or lose contact to the other boats. This is especially true in tight formations. What's more, when a line is attacked, not all boats are equally close to the victim and some may take a long time to reach the scene and provide assistance.

The 'front' formation puts all ships on a line next to each other. This is very difficult to maintain during longer passages, and it is not easy to keep your lateral distances as well as a good 'front line' with small crews. In tight formations, some boats may reduce the performance for those sailing downwind as they disturb the wind. However, the front is quite an effective formation to be used in a confrontation, so a convoy could consider taking this formation under engines to close in on any suspicious target. To present a compact formation, the distance between convoy boats should be as close as skippers consider safe. The bows of three vessels approaching in this fashion may very well intimidate the less-professional attacker in a small fishing boat or skiff.

Convoy tactics: reacting to suspicious vessels

The reaction to a suspicious approach needs to be discussed on two levels, namely regarding convoy tactics as well as the individual boat attacked by the assailant.

In this section, we will focus on the convoy level. Details of how the crew may react to a suspicious approach towards their single boat have already been covered. The general reaction schemes as described in Fig 4.17 apply to convoys too.

Spotting a suspicious vessel

The moment that a convoy member spots a suspicious vessel, he should determine the vessel's distance to the nearest convoy ship, along with the bearing and speed of the potential threat. He will then inform all other convoy ships and share this information.

If the threat has been spotted close to the convoy, all boats should react as described in the 'offensive approach' section below.

Whenever the spotter has detected the vessel far away and the vessel is not directly approaching the convoy, all boats should put eyes on it and monitor its actions. Any convoy ships that are not readied for a potential boarding should now finalise all preparations.

All crew members of all ships should be mustered on deck. The crews should prepare to drop sails at an instant (if weather and direction permits a faster cruise under sails, this step may be omitted) to either dash for assistance or run towards safety without any restraints. The engines should be started now. All items demonstrating strength (weapons, replicas, flare guns, etc) should be readied and presented.

If the suspicious vessel does not approach or it proves to be clearly harmless, convoy routine will be reinitiated as soon as the vessel is behind the horizon. It is strongly recommend that you keep eyes on the vessel until it drops behind the horizon.

Whenever the vessel is identified as a pirate vessel or keeps approaching the convoy, crews should move to the next level: deterrence and defence.

Offensive approach: deterring and fending off pirates

Just as solo cruisers would, each boat will determine whether they could run at least at the same speed as the approaching vessel. In this situation, the joint decision on mutual assistance will also be called to mind, helping the crew of each boat to decide how to react. Accordingly, convoy boats will attempt to escape together, approach the attacker together in a front formation, or decide to end convoy operations and act as single entities following their individual abilities and preferences.

If boats do not wish (or are unable) to run, a show of force or at least a pretence of it may be an option against a non-heavily-armed assailant. In the following paragraphs, we assume that the convoy will try to fend off the attackers together.

174 THE COMPLETE YACHT SECURITY HANDBOOK

▸ Convoy-Vessel   ▸ Suspicious Vessel

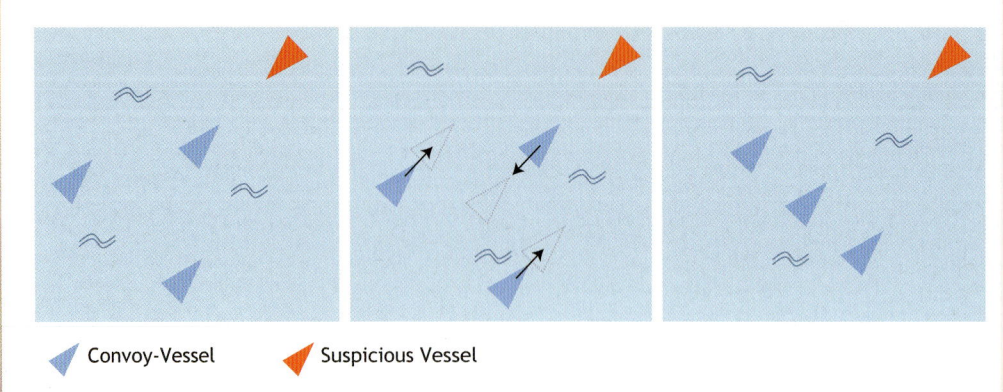

FIG 4.25: Changing from 'triangle' to 'front' formation when approached from ahead.

▸ Convoy-Vessel   ▸ Suspicious Vessel

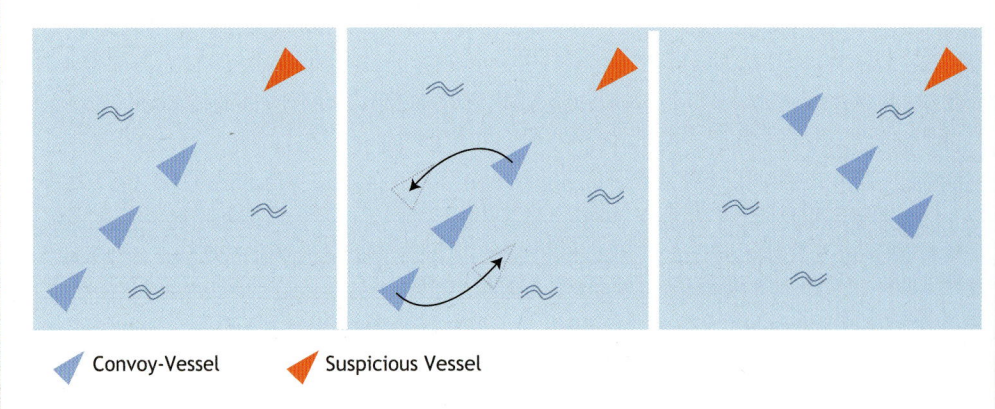

FIG 4.26: Changing from 'line' to 'front' formation when approached from ahead.

▸ Convoy-Vessel   ▸ Suspicious Vessel

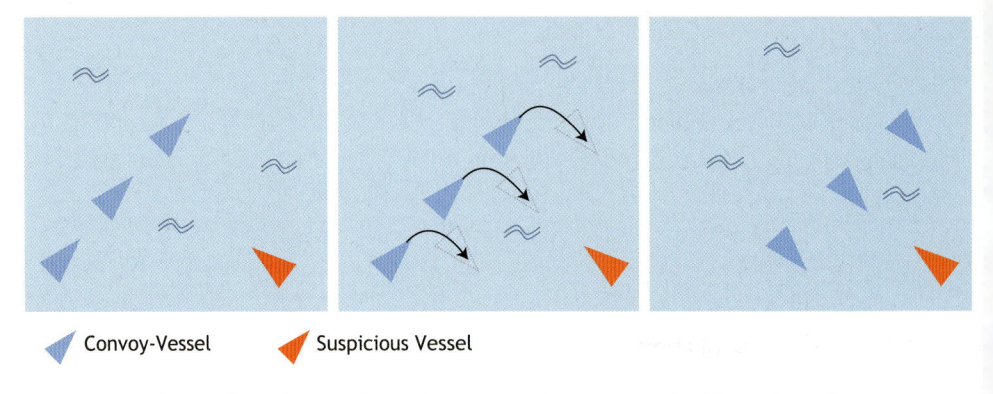

FIG 4.27: Changing from 'line' to 'front' formation when approached from the side.

In a first step, all sails that reduce manoeuvrability under the current circumstances are dropped or fixed in a way that allows the boat to choose its course freely. The engines should be running at this moment as most of the following manoeuvres are conducted under engine power.

If there is sufficient time, the convoy will move to the 'front' formation and approach the attacker directly. The boats should try to maintain a distance of no more than two boat lengths between them. Many small-time pirates would rather abort their approach than risk a collision with a convoy showing such determination.

If the vessel approaches the convoy from ahead while it is in a 'triangle' formation, the leading boat should simply reduce speed while the others catch up. See Fig. 4.25.

If the suspicious vessel approaches the convoy from ahead while it is in a 'line' formation, the leading vessel drops back and the last convoy ship catches up. Both take a lateral position next to the boat in the middle. See Fig. 4.26.

If a 'line' formation is approached from either side, the boats would simply turn towards the threat. See Fig. 4.27.

These manoeuvres can easily be modified to meet an approach from any direction.

When executing your moves, no one expects skippers and crews of pleasure boats to win a beauty contest. However, it may be a good idea to practise them once or twice prior to your cruise or soon after you have embarked as a convoy.

If your convoy does not have the time to get into a 'front' formation, each boat should nonetheless close the gap to the attacked boat as quickly as possible to show that they are acting together. It would be good to approach in close formation: as for the 'front' formation, it is best to keep a distance of no more than two boat lengths.

One of the boats further away from the attacker should issue warnings at him. Ideally, the message should be issued in the local language. Alternatively, English is acceptable, although there is a risk that the message will not be understood. As described earlier, be very clear and offensive in this communication, leave no room for misunderstandings open, and threaten that you will use arms if there is a further approach. Whether you really have firearms on board any convoy ship is not important at this point. Have all crew mustered on deck in this phase. If available, present firearms or replicas to the approaching vessel. Use hand and acoustic signals to order the vessel away from your convoy.

If the approach continues until it is closer than about 200 metres (yards), the group should start to fire warning shots from firearms or signalling devices across the vessel's bow. When engaging an offensive vessel in this manner, act in a determined manner but keep the rules of good measure in mind. How to properly deter without risking the lives of innocent mariners is described in the '50-metre perimeter' section of the 'Crew tactics: avoidance and defence' section (see page 145).

If the approach is not aborted at this point, you can be certain that an attack is imminent. Now, the division of labour commences. The boat directly approached will focus on fending off the attacker and any boarding attempts while the other boats communicate with potential support and deploy emergency signals. Depending on their ability, some or all of the boats could decide to directly support the attacked ship.

However, deploying emergency signals is a priority. All available one- and two-way signals should be used: MAYDAY called via distress buttons on HF and UHV radios as well as satellite phones, EPIRBs and PLBs. If available for your region, call the contact numbers issued by the responsible coast guard, navy or police. Flares and parachute rockets should definitely be deployed during night and day.

After all emergency communication has finished, crews could decide to support the boat under direct attack to prevent any boarding. Depending on their availability, firearms, flare guns, parachute rockets etc could be used to distract the boarding vessel and put pressure on its crew. Depending on the relative size of the pirates' vessel, ramming can be an option for a sturdy sailor against a skimpy skiff.

Always remember that such actions against pirates who are heavily armed with multiple assault rifles are extremely dangerous and should not be tried by cruising crews. It is important that the attackers' weapons are not capable of severely damaging or destroying the supporting convoy ships.

Convoy actions to support vessels about to be boarded

If the convoy ships are sturdily built, well equipped and staffed by a determined as well as capable crew, they may decide to directly support their convoy mate while fending off boarders.

When pirates are trying to board from smaller fishing craft or even skiffs, the act of boarding puts them into a very weak situation. A relieving assault from one or two yachts can seriously interrupt such an operation and destroy the attackers' motivation.

There are several options when it comes to supporting your convoy mate in a direct conflict. If there are firearms on board, the pirates could be engaged from a distance. This may be targeted at the attacking vessel before a contact for boarding is made. At a later stage, it could be aimed to suppress the threat that may be coming from the pirate crew remaining in their vessel while others attempt to board the target. The boarders would now lack their fire support and the attacked crew would have a better chance, with less risk, of repelling the pirates' attempts to step over. To achieve this, there is not necessarily the need for a wild shoot-out: some well-placed shots will usually result in a distraction from the boarding action and should at least succeed in keeping some heads down, taking some pressure off the crew in the boat to be boarded.

Remember that engaging in a firefight is a dangerous option for crews who are not trained in combat tactics. They risk their lives from live fire returned by practiced criminals or terrorists.

If the attackers are well armed and determined, it might be best not to target the crew directly. Instead, in most non-professional pirate attacks, you will have the option to aim at their rudders, engines or other vital parts of their vessel. In many cases, this will keep their heads down without you risking seriously injuring or killing any person.

If there are no firearms on board the convoy vessels and the attackers are not armed with rifles, you could attempt to distract the attackers by using improvised weapons. Flares and parachute rockets could also be used to disrupt pirate actions and seriously endanger any vessel made from fibreglass or wood. Finally, there is always the option of ramming a weak vessel with a sturdy yacht.

Remember to continuously call for support on all available emergency channels while engaging actively. A swift support mission by trained security forces is still your best option in order to survive the encounter unharmed.

Convoy operations after a boarding

As stated in the sections covering single-boat actions, most crews who are not trained and equipped for combat will surrender the moment a party armed with guns has successfully boarded the boat.

In such an event, it is advisable that the other convoy boats retreat to a position at least 220 metres (yards) away from the boarded vessel. This will make it very difficult for pirates to hit them using small arms fire and they can still assist from a distance.

The simple presence of other boats will have a good effect on the morale of the surrendered crew and will sustain pressure on the boarders, too. However, it is important that each crew has shared their view on how the others should react in such a situation. It may be possible that pirates threaten to hurt their hostages if the other boats do not retreat. All crews should have shared opinions in advance concerning how they wish their mates to act under such circumstances.

If boats remain close, they should try to collect as much information on the attackers as possible. How many are on board? What is their armament? Are pirates or crew members injured? Is there an identified leader? What is their watch routine? Is the pirate vessel still operating? This information is very valuable if and when professional rescue parties arrive on the scene.

If the remaining boats are equipped with long-range communication equipment, they should try to establish and maintain continuous communication with local coast guards, MRCCs or military units. The earlier such units arrive and the better their picture of the scene, the more likely it is that the hostages will be freed unharmed.

Aside from surveillance and communication, boats remaining on the scene could also support the attacked vessel with food, water and emergency equipment. They will also be able to pick up any crew who go overboard or assist in medical emergencies.

SECURITY TEAMS ON BOARD

Security contractors specialising in the protection of naval vessels operate along the fringes of most high-risk areas for commercial shipping. At the time of writing, these are focused on the northern Indian Ocean/Gulf of Aden as well as the Gulf of Guinea.

Other areas that are dangerous for yachts but hold low relevance for commercial shipping could be covered but are usually not within the standard portfolio, making an already-expensive service even less affordable.

Due to the relatively high costs of such arrangements, simply avoiding the dangerous area might be a better strategy for yacht crews. Unfortunately, though, there are some bottlenecks in global shipping routes that can only be avoided at high costs. Take the Gulf of Aden/Suez passage: if you wanted to avoid that route on your way from Asia to Europe, you would have to take a detour via Cape of Good Hope.

The security companies' services normally involve adding armed guards to your vessel. The operators will usually be retired soldiers with a navy/marine or special-forces background. When contracting them, skippers add crew with training (and sometimes experience) in armed confrontation. They also add firearms (usually semi-automatic or assault rifles), which are a standard equipment of the operators.

As most countries allow neither armed personnel nor military-style guns on their territory, arms are usually taken on board from small freighters just inside international waters close to the port of departure or en route close to the major shipping routes.

Most contractors will insist upon putting at least three guards on your vessel. The primary reason for this is that it gives them the ability to maintain a continuous vigilant watch. Each heavily armed guard will also increase the general firepower and thus the chances of prevailing in a conflict.

However, the yacht needs to have room for the extra crew and their bulky gear. The crew also needs to accept the presence of the added personnel on board, who cannot be selected on the basis of their character. The vast majority of reports about the operators' behaviour on board yachts are positive, although you simply do not know with whom you will share the narrow space of your vessel for the weeks of your passage.

Expenditure can vary between contractors. Often it is per detachment, per day. Calculating on a daily basis leads some skippers to think even harder about the fuel that they take on board before embarking for a passage through pirate-plagued waters with guards. Each day of being lost in calm conditions results not only in an added risk of being discovered but also an added cost for the security team.

A high-quality contractor will usually arrange for the necessary formalities and operations for his crew. This includes embarking, the pick-up and drop-off of guns, and eventual immigration formalities if the guards are not directly handed over to the next ship. Nevertheless, skippers need to calculate for extra time, potential detours and extra food and drink.

ARMED ESCORTS

Armed escorts are the luxury version of armed security teams on board. Rather than having the operators lodging on board your yacht, they will guard your vessel or a convoy from an accompanying boat.

While adding extra security and comfort to the set-up, this option also adds the charter and fuel price to the total security bill. It is extremely expensive.

The concept emerged at a time when armed guards were not allowed on civil or merchant ships (pre-2011). As this has since changed, the service has become rare. Given that the total price for such a passage will get close to the costs of transporting a yacht through a high-risk zone on a freighter, it is not elaborated here.

ALTERNATIVE: TRANSPORT ON A FREIGHTLINER

Nowadays, transport through high-risk bottlenecks can also be arranged on a freightliner. As for security contractors, this strategy only makes sense if there is no reasonable alternative route that means you can avoid the high-risk passage. Depending on the developments in Somalia, the Gulf of Aden would qualify today. Many years ago, the Strait of Malacca would have been a candidate.

Although the first quote is usually shockingly high, skippers need to take a second look when comparing the price with a normal passage. How much fuel, wear and tear did you calculate for your passage? Adding an armed security team makes the passage even more expensive, so perhaps using a freightliner isn't so bad after all.

You can find the largest contractors such as Sevenstar Yacht Transport or DYT Yacht Transport on the internet and contact them to quickly get a quote.

Prices greatly vary depending on the size of the yacht, the travelled route and the time of year. To give you some examples (prices will change, so do look them up yourself), longer hauls such as the passage from Asia to the Mediterranean were priced at about US$45,000 for a 54ft sailboat in the past, and it cost US$47,000 USD to transport a 40ft catamaran from Phuket to Turkey in early 2017.

Skippers should be careful when selecting their carrier. The marine logistics business is in a long-lasting phase of low business and over-capacity. One of the large companies, Yacht Path, went bankrupt some years ago. Their assets were acquired by another company, which now offers their services as United Yacht Transport. Customers who had paid in advance had problems getting their money back for blown deals, while others reported long processes to get their boats returned.

While such economic disasters are exceptional, there are quite a few reports of impromptu changes to shipping schedules or passages that are simply cancelled just before the planned departure dates. Therefore, it is wise to negotiate the procedure for delays or cancellations in advance.

A limiting factor for this strategy is the ports of operation. At the time of writing, only Sevenstar services the Asia-Med route whereas DYT focuses on the Americas.

FIG 4.28: *A very secure albeit expensive option to bypass pirate-plagued waters. (Photo: Sevenstar)*

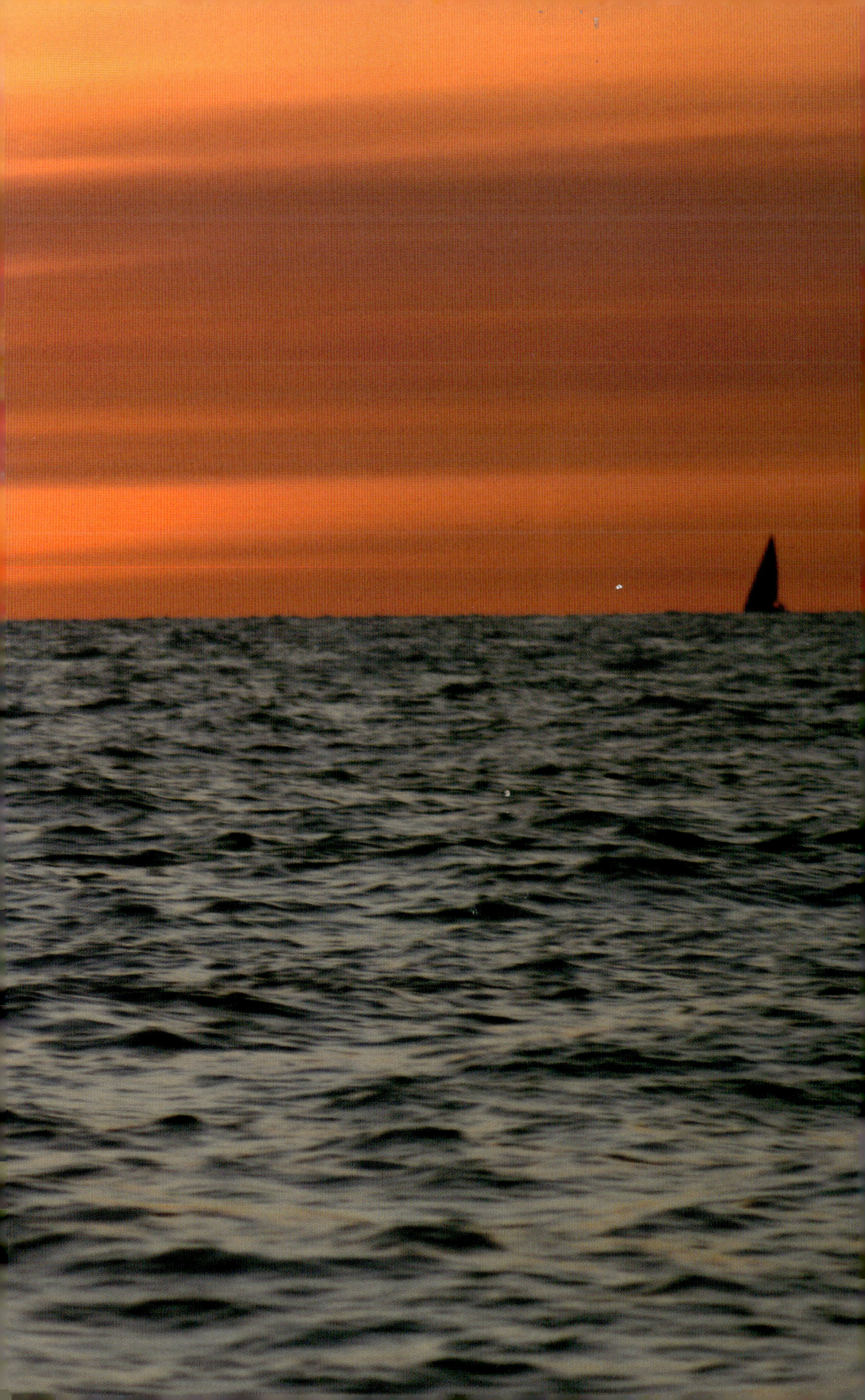

APPENDICES

ITEMS OF A CRIME REPORT

- Type of attacked vessel
- Yacht situation (in port, moored, at anchor, at sea)
- Location of the attack
- Number of crew on board at time of attack
- Other boats nearby and approximate distance

- Date and time of attack
- Method of approach
- Number of attackers
- Point of entry or boarding
- Armament carried
- Duration of attack
- Attacker description
- Method of action (stealth or surprise, deception, threats, violence)

- Crew's action (e.g. alarm raised, barricade, resistance, surrender)
- Attacker's reaction to crew
- Support by other boats, coastal community or officials during the attack
- Crew injured (number and type of injuries)
- Attackers injured
- Attacker's method of leaving vessel
- Crime success / failure
- Type of booty (if any)

- Type of gear for detection, deterrence on board
- Boat locked or open
- Method to hide / secure stolen or robbed items

- Activity after attack (remain on site vs departure)
- Involvement of officials and their reaction

EMERGENCY COMMUNICATION TABLE
to be printed and located at VHF / sat phone (example from Gulf of Aden)

PROCEDURE IN CASE OF ATTACK:

1. Push Distress Button on VHF for 5 Seconds

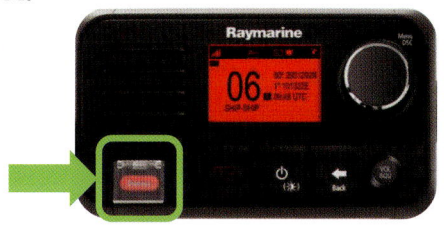

2. CALL UKMTO on sat-phone
Speed Dial: 111
Regular dial: +44 2392 222060
Have position at hand

Our Vessel
Name: STINGRAY
MMSI: 123456789
Callsign: HFG123

Our position data found on VHF display

VHF VOICE MESSAGES
STAY AWAY call (VHF channel 16); repeat twice

"Unknown vessel, unknown vessel.
You are approaching the armed vessel "*Stingray*".
Abort your approach immediately or our security team will engage you.
I repeat: abort your approach immediately or you will be shot at. Go away! Now!"

MAYDAY call (VHF channel 16); repeat twice

Mayday Mayday Mayday
This is yacht "*Stingray*", MMSI 123456789
On position
xx degrees, yy decimal, zz minutes north
xx degrees, yy decimal, zz minutes east
We are under pirate attack and require urgent assistance

GLOSSARY

Automatic
In terms of firearms, any gun that fires rounds continuously as long as the trigger is pressed. They do not tend to be accurate or efficient, especially over longer ranges, but are effective as an offensive weapon because the rate of fire maximises the probability of the attacker hitting a target in the short to medium range. See also: *semi-automatic*.

Bolt-action
A type of firearm (more often rifles than shotguns or handguns) where rounds of ammunition are loaded in and out of the barrel chamber by manually operating a bolt handle. It has better precision than automatic and semi-automatic weapons, but is slow to reload, so is mostly used as a hunting weapon.

COLREGs
The International Regulations for Preventing Collisions at Sea, published by the International Maritime Organization, covering the 'rules of the road' and navigation conventions, such as factors to consider when determining who should have right of way.

EPIRB
Emergency Position Indicating Radio Beacon. An emergency device that uses a radio signal to indicate the position of an individual (see also: *PLB*), vessel or aircraft in trouble. Distress alerts are picked up by Cospas-Sarsat satellites used for search and rescue (see: *SAR*). Some require manual activation but many used on boats are activated when they come into contact with water. Modern EPIRBs with GPS allow the beacon to be accurately tracked to within 100 metres (330 feet).

FOM
Figure of merit. In relation to night vision goggles, the FOM indicates the quality of the image, graded according to the system's noise. The higher the FOM, the better the quality, with anything with a FOM of 1,600 or higher being useful at sea.

GPS
Global Positioning System. A system of sattellites that allow a boat's position to be calculated with great accuracy (within 5 metres, or 16 feet) using an electronic receiver.

GRIB
GRIdded Binary or General Regularly-distributed Information in Binary form. A concise data format commonly used to transmit weather forecast information. The data packets tend to be very small, so are easily downloaded.

GSM
Global System for Mobile communication. The standard for cellular communication networks (mobile phones, in contrast to satellite phones).

HF
High Frequency. Because radio waves in this frequency range bounce off the atmosphere, they are suitable for long-distance (including intercontinental) communication, so are used in aviation and by weather stations. See also: *VHF*.

IP rating
International Protection (or Ingress Protection) code, classifying the degree of protection against intrusion (such as by water) of mechanical casings and electrical enclosures.

Lever-action
A type of firearm (mostly rifles) with a lever located near the trigger to load fresh rounds into the barrel chamber. Lever-action weapons have limited military use, but remain reasonably popular in sport shooting and hunting for sharing much of the accuracy of bolt-action (see above) weapons but with faster reload times.

MARPA
Mini Automatic Radar Plotting Aid. A radar feature that tracks specific targets, primarily with the goal of avoiding collision. Once a target is manually selected, its range, bearing, speed, direction and the amount of time it will take to reach the tracking vessel are automatically calculated and updated, which can prove useful for monitoring suspicious boats.

PLB
Personal Locator Beacon. A type of EPIRB (see above) registered to an individual rather than a boat. It is also usually smaller and has a much shorter battery life, but unlike an EPIRB, which may float free of a vessel, a PLB can help locate someone who has been separated from their boat.

Pump-action

A type of firearm (shotguns and rifles) with a sliding fore-end, which can be pumped to manually eject a fired round and load the next. They are slow to reload, as ammunition has to be inserted individually. See also: *bolt-action* and *lever-action*.

SAR

Search and Rescue. In this context, air-sea rescue, typically involving fast response vessels, seaplanes, helicopters, both military and civilian, searching for and recovering sailors and passengers of sea vessels in distress, or aircraft downed at sea.

Semi-automatic

In contrast to automatic firearms (see above), semi-automatic weapons fire only one round with each trigger pull (though the next round is automatically reloaded, as with fully automatic weapons). While more accurate than automatic weapons over medium range, the slower reload rate lowers the probability of hitting the target.

VHF

Very High Frequency. VHF radios are the most common type carried on boats. VHF radio waves have a range of abut 60 nautical miles (110km) between boats at sea and tall aerials on land, down to 5 nautical miles (9km) between boats at sea. See also: *HF*.

ABOUT THE AUTHOR

Fritze von Berswordt, born in 1972, is a psychologist, strategy consultant and circumnavigator.

Some 20 years ago, while travelling off the beaten track in Africa; the Middle East, South East Asia, Latin and South America, he started thinking about safety. Security as a means to a single cause: to explore remote countries and cultures.

Driven by his passion for discovery and love of travelling, Fritze lived a sailor's dream together with his wife and daughter: a circumnavigation within two years. Curiosity led the family away from established routes to many regions that are generally given a wide berth by the cruising community since they are deemed unsafe.

Unforgettable experiences on the Solomon Islands, in Papua New Guinea, in West Papua, the north of Sri Lanka, Sudan and the 'high-risk area' (HRA) of the Indian Ocean were their rewards for having made the effort to research, prepare and take well-calculated risks.

In addition to his work as a strategy consultant, Fritze is member of the executive board of the 'Stiftung Situation Kunst', a foundation that not only runs a museum but also strives to support the arts, science and education. In his spare time, he is a sailor, traveller, bee keeper and hunter.

REFERENCE

Bateman, S, 'Piracy and armed robbery against ships in Indonesian waters', in Cribb, R, *Indonesia beyond the waters's edge: managing an archipelagic state,* Singapore: ISEAS Publishing (2009), pp117–133.

Bateman, S, *The true story of piracy in Asia, A closer look at the statistics shows that piracy in Asia is not the problem it is made out to be* (4 April 2016), retrieved from Asia & the Pacific Policy Society: www.policyforum.net/the-true-story-of-piracy-in-asia/

Beckmann, RC, 'Combatting Piracy and Armed Robbery Against Ships in Southeast Asia: The Way Forward', *Ocean Development & International Law* (2002), pp317–341.

Block, R 'Victim-Offender Dynamics in Violent Crime', *Journal of Criminal Law and Criminology, 72* (Summer) (1981).

BMP4 Best Management Practices for Protection against Somalia Based Pirac, Edinburgh: Witherby Publishing Group (2011).

Fajnzylber, P, 'What causes violent crime?', *European Economic Review* (2002), pp1323–1357.

Fiorelli, M, *Piracy in Africa: The case of the Gulf of Guinea,* Accra: Kovi Annan International Peacekeeping Training Centre (2014).

Groff, E, *Modeling the Dynamics of Street Robberies,* Temple University, Institute for Law and Justice, Alexandria, Virginia: Office of Justice Programs, National Institute of Justice (2008).

Hardberger, M, *Seized!* London: Nicholas Brealey Publishing (2010).

Heinonen, JA, *Home Invasion Robbery,* Washington: Office of Community Oriented Policing Services (2012).

ICC International Maritime Bureau, *Piracy and armed robbery against ships, Report for the Period of 1 January – 30 June 2016,* London: ICC International Maritime Bureau (2016).

Indermaur, D, 'Reducing the opportunities for violence in robbery and property crime: The perspective of offenders and vicitms', in Homel, R, *Crime Prevention Studies,* Monsey: Criminal Justice Press (1996), pp133–157.

Kashuba, SD, *Resistance of exterior walls to high velocity projectiles,* Canadian Police Research Center (2002).

Lindegaard, MR, 'Consequences of Expected and Observed Victim Resistance for Offender Violence during Robbery Events', *Journal of Research in Crime and Delinquency, 52(I)*, (2015), pp32–61.

Maguire, MM, *The Oxford Handbook of Criminology*, Oxford: Oxford University Press (2012).

Matthews, R, *Armed Robbery*, New York: Routledge (2012).

Morrison, SA, 'An analysis of the decision-making practices of armed robbers', in Homel, R, *Crime Prevention Studies* (pp159–188), Monsey: Criminal Justice Press (1996).

Neumayer, E, 'Is inequality really a major cause of violent crime? Evidence from a cross-national panel of robbery and violent theft rates', *Journal of peace research* (2005), pp101–112.

www.noonsite.com - Piracy and Security (2016), retrieved from www.noonsite.com/General/Piracy

Oceans Beyond Piracy (10 December 2015), *The state of maritime piracy 2015*, retrieved from 'Piracy and robbery against ships in the Gulf of Guinea 2015' http://oceansbeyondpiracy.org/reports/sop2015/west-africa

Oceans Beyond Piracy (2016), *Gulf of Guinea Trend 2016: kidnap for ransom piracy trending higher in Gulf of Guinea*, Broomfield: Oceans Beyond Piracy.

O'Donnell, I, 'Armed and dangerous? The use of firearms in robbery', *The Howard Journal of Criminal Justice* (1997), pp305–320.

Oliveira, CA, 'The impact of private precautions on home burglary and robbery in Brazil' (15 July 2014), *SSRN*, Cristiano Aguiar de Oliveira, retrieved from https://ssrn.com/abstract=2466590

Osinowo, AA, *Combating Piracy in the Gulf of Guinea*, Washington: Africa Center for Strategic Studies (2015).

Richardson, L, *What Terrorists Want*, Random House (2006).

Steffen, D, *Quantifying Piracy Trends in the Gulf of Guinea – Who's Right and Who's Wrong?* (19 June 2015), retrieved from USNI News: https://news.usni.org/2015/06/19/essay-quantifying-piracy-trends-in-the-gulf-of-guinea-whos-right-and-whos-wrong

Storey, I, 'Addressing the persistent problem of piracy and sea robbery in southeast asia', *Institute of South-East Asian Studies – Perspective* (2016).

Suarez, G, *The Tactical Advantage*, Boulder, Colorado: Paladin Press (1998).

Yapp, JR, *The profiling of robbery offenders*, Birmingham: The Centre for Forensic and Criminological Psychology (2010).

INDEX

abduction 16
   deterrence of 19–20
acoustic security devices 79–80
alarms 71–87
ambushes
   avoiding 147–53
   dealing with 141–2, 158–63
   deterrence of 153–8
anchorages
   assessing risks 113–16
   confronting boarders 121–7
   departures (emergency) 127
   leaving the boat 127–9
   nighttime 119–21
   visits by the authorities 129–30
anchor chains, security concerns of 62–3
attackers
   influence of drugs 33–4
   types of 28–34, 132–3
   use of firearms 43, 44
attacks
   controlling 12–13, 19–23
   defence and resistance against 87–8
   defensive tactics 23–6
   dynamics of 17–28
   silent 12–13, 18–19, 21–3
   types of 12–17, 21
   types of attackers 28–34
   weapons involved 22, 43, 44
barbed wire 90–1
barriers and blocking devices 88–91
bathing platforms, risks of 51–2
bays, solitary 112–30
bearing compass 69
binoculars 67–8
bladed weapons 98–100
boardings
   dealing with 121–7, 144, 159–63, 176–7
   methods 50–2
boat layout 50–5
Brazil 23
burglary 14–15
   deterrence of 18–19, 63–7
   in marinas 111
Caribbean 11, 23, 39, 44, 47, 134, 144
CCTV cameras 84–5
communication
   convoys 168–70
   Digital Selective Calling 71–4
   HF radio 73–4
   satellite phones 74–5
   VHF radio 71–2
companionway security 63–7
compass, bearing 69
construction (boat), security concerns of 52–4
controlling attacks 12–13, 19–23
convoys
   boardings by pirates, response to 176–7
   communication and coordination 168–70
   formations 170–2
   goals of 165–7
   lookouts 170–1
   preparation 164–77
   pros and cons 163–4
   tactics for deterring and fending off pirates 173–6
   watchkeeping 170–1
course-plotting in pirate-plagued waters 136–8
crew behaviour and capabilities 101–5
crime reports 182
decklights (as deterrence) 80–1
defensive tactics 23–6, 87–8
detecting attacks 67–71
deterrence 71–87

INDEX

abduction 19–20
burglary 18–19
robbery 19–20
theft 18–19
Digital Selective Calling 71–4
dinghy protection 58–63
distress signals
  procedure 183
  pyrotechnic 76–7
Dominican Republic 41
drugs, influence on attackers 33–4
dynamics of criminal attacks 17–28
dynamics of pirate attacks (offshore) 140–63
electric fences 91
EPIRBs 76
escorts, armed 179
firearms 91–6
  legal situation 91–2
  replica 81–2
  types of 93–6
  use by attackers 43, 44
flare guns 77–9
flashlights, high-powered 86–7
freeboards, security concerns of 50–2
Guatemala 23, 43, 132, 170
Gulf of Aden 56, 72, 73, 74, 132, 133, 149, 178
Gulf of Guinea 39, 43, 131–2, 133, 143, 144
handguns 93–4
hatch security 66–7

headlamps 100–1
HF radio 73–4
Honduras 23, 41, 42–3, 132, 170
hull shapes, security concerns of 50–2
injuries resulting from attempted resistance 24–5, 27
knives 98–9
lasers
  signalling 79
  as weapons 96–7
layout
  of boats 50–5
  of marinas 109–10
locks 57–8
lookouts 67–71 150–2, 170–1
mace 97–8
machetes 98–9
Malacca, Strait of 11, 31, 72, 133, 149, 160–1, 169
Malaysia 11
marinas
  assessing risks 108–9, 110–11
  burglary in 111
  design and security 109–10
  theft in 111
mooring buoys 112–30
murder 16
Nigeria 42–3, 143
night vision goggles 69–70
outboard protection 58–63
Panama 23, 43, 110

pepper spray 97–8
Philippines 23, 32–3, 39, 42–3, 47, 132, 137
piracy 130–63
  assessing risks 133–5
  avoidance and defence 135–40, 145–60
  crew preparation 139–40
  dynamics of offshore attacks 140–63
  nature of the threat 10–11
  preparation before cruise 133–40
  reaction to resistance 31–2
  strategies in pirate-plagued waters 135–40
  tactics of pirates 141–5
  types of attackers 132–3
  yacht preparation 138–9
planning a cruise 36–47
PLBs 76
police
  pirates disguised as 129–30
  rescue by 160–3
preparation before a cruise 48–105
  sailing in pirate-plagued waters 133–40
propeller-fouling devices 87–8
radios
  HF 73–4
  VHF 71–2
razor wire 90–1

rescue by military or police 160–3
resistance vs submission 23–8
rifles 94–5
risks
    assessing 39–45
    reducing 46–7
robbery 15
    deterrence of 19–20, 63–7
routing through pirate-plagued waters 136–8
satellite phones 74–5
security equipment 55–101
    alarms 71–87
    barriers and blocking devices 88–91
    companionway security 63–7
    defence and resistance against attacks 87–8
    detecting attacks 67–71
    deterring attacks 71–87
    dinghy protection 58–63
    hatch security 66–7

outboard protection 58–63
security teams (having on board) 178
self-defence
    reasonable levels of 35
    risk of injuries 24–5, 27
shotguns 95–6
signs, warning 85–6
silent attacks 12–13, 18–19, 21–3
Singapore 11, 109
solitary bays 112–30
Somalia 42–3, 132, 143
Strait of Malacca 11, 31, 72, 133, 149, 160–1, 169
submission vs resistance 23–8
tear gas 97–8
terrorism 17, 32–3
theft 13–14
    deterrence of 18–19, 57–63
    in marinas 111
thermal imaging equipment 70–1

transportation on freightliner 179–80
types of criminal attack 12–17, 21
    abduction 16
    burglary 14–15
    murder 16
    robbery 15
    terrorism 17, 32–3
    theft 13–14
    war, acts of 17
Venezuela 23, 39, 42–3, 132, 134, 171
VHF radio 71–2
war, acts of 17
watchkeeping 67–71 150–2, 170–1
weapons
    attacks involving 22, 43
    firearms 91–6
    knives 99–100
    lasers 96–7
    mace 97–8
    machetes 98–9
    pepper spray 97–8
Yemen 42–3